**THE BMW CLOTHING RANGE.
NEEDLESS TO SAY IT HANGS BEAUTIFULLY.**

Shaped from Schoeller K-300, a material exclusive to BMW and five times stronger than steel, the BMW Madison jacket and the complete 1994 Clothing Range can be picked up at your BMW Motorcycle Dealer. Or call 0800 325600 for a brochure. **THE 1994 CLOTHING RANGE**

THE BIKER'S BIBLE

Windrow
&
Greene PUBLISHING

THE BIKER'S BIBLE

Windrow & Greene's

Motorcycle Enthusiast's

Directory & Sourcebook

© 1994 Windrow & Greene Ltd.

This edition first published in
Great Britain 1994 by
Windrow & Greene Ltd.
19A Floral Street
London WC2E 9DS

A CIP record for this book is
available from the British Library

ISBN 1 872004 49 0

Printed and bound in Hong Kong by
World Print Ltd.

Advertising agents for this book:
Boland Advertising Ltd.
Salatin House, 19 Cedar Road
Sutton, Surrey SM2 5JG
Tel: 081 770 9444

Designed and produced by
ghk DESIGN
10 Barley Mow Passage
London W4 4PH
Tel: 081 994 7054

Front cover illustation:
*Reigning World Superbike Champion
Scott Russell puts the new Kawasaki
Ninja*[TM] *ZX-9R through its paces.*

CONTENTS

Top Ten Tackle

*From Sportsters to Tourers
every Sector has its Winner*

Our thoughts on ten for 1994 follow,
in strictly alphabetical order...

*Bimota YB9 SR
(Jason Critchell)*

(1)   Bimota YB9 SR

(Jason Critchell)

For a company that only made about 600 bikes last year Bimota has an amazingly high profile, and an equally amazing list of engineering firsts. The little firm near Rimini has been making frames for bikes since long before Johnny Cecotto won the 350cc World Championship back in 1973 with a Bimota frame. After another four years came the SB2, the first road bike, which carried the Suzuki GS750 engine; and after that came a flow of frames and chassis which tamed the huge power outputs of the big Japanese engines, at a time when many riders less wealthy had to make do with wallows and wobbles of terrifying frequency.

The Ducati-powered DB1 and DB2 were both visually stunning, and have led to a whole host of imitators even from the major oriental factories; but, from a business point of view, an equally important event came in 1986 when Bimota managed to secure a deal with Yamaha to buy the engines direct from the factory. Up till then they had to buy complete bikes and break them for the motors, but now the economies of scale changed the company's fortunes.

Eight years further on we get the newest Bimota, aimed not at the super-rich but at a target audience that might be able to afford some sort of Bimota even after years of world-wide recession. This is, of course, relative since you still need ten grand to buy the YB9 SR, with its FZR600 motor. Even though this is a relatively affordable Bimota, it is still a beautifully crafted piece of kit. The bodywork, something of a trademark, is swoopy and curvaceous, with a passing nod at the need to take a pillion.

The bodywork curves over a massive beam frame, alloy of course, with conventional Paioli front forks instead of the hub-centre steered Tesi system. The front end is relatively soft and pliable without letting any wandering or weaving set in. It needs a good front end, since the rider is putting a fair bit of weight on it; the YB9 SR has an incredibly short wheelbase at 1380mm. There is a reasonable stretch to the bars, but the distance from seat to footrest puts this machine in the category of a race-replica rather than a sportster with any all-round pretensions.

At just 175kg, you can flick this bike around like a whip, the steering being noticeably faster than on a stock and far-from-slow FZR600. The engine suits this perfectly, being stock apart from intake and exhausts, which allow a free and faster flow of gases. You need to rev it; there are no surprises there, but spinning the motor gives you a huge surge of power. Keeping it above 8000rpm gives you the ride you crave, while the six-speed box keeps things slickly in tune. This is the sort of little bike that appears in much bigger bikes' mirrors and then blows them away. It's a real giant-killer.

And when things get fraught the Brembo Goldlines can nail the front end down, firmly enough to make your head bang into the rather tight screen. It goes, it handles, it stops. What more could you want? Oh yes, of course, you want quality. Well, you get a gorgeous aluminium frame, immaculate welding, parts milled from billet, and the feeling that every single little item has been carefully thought through. Delightful details abound, like the four fasteners that hold on the entire seat and tank cover. This is more than designer chic, this is quality in depth.

What else could you desire? Exclusivity? You got it. There will only be 18 models of this Bimota brought to the British shores in 1994, which makes it kind of surprising that it is only £9999, since the price would make you think they intended to bring in many more. The 18 owners probably won't be complaining.

Price:	£9999
Power:	98bhp at 11,500rpm
Top speed:	140mph

(2) BMW R1100RS

BMW is not a company that one instantly thinks of as a sparkling innovator on the scene, more as a bunch of reactionary codgers. Yet it was BMW that first introduced a wind tunnel-tested fairing on a production bike; they were the first to bring in ABS, and the first to put a monoshock rear on a bike — others may have done it earlier but BMW were the first to do it on mass-production scale. Similarly, the company did show signs of finally losing interest in the flat twin design which it had championed since 1932, although public demand — that fickle mistress — ensured its continuance in a world very different to that into which it was launched. Now, apparently several centuries later, we have four-valve heads — but, just as importantly, we also have a weird front end on a production bike. What a strange company…

The Telelever front end complements the Paralever rear, giving you two springs to cope with the bike's not inconsiderable 239kg. The Telelever means that the engine becomes a stressed member, with a Showa spring mounted centrally behind the fairing, with the horizontal control arm pivoting off the crankcases themselves at the rear, and off the "forks" at the front. The forks themselves are not telescopic; the suspension movement is taken by the control arm and the top of the steering stem. With this system you can dial in any amount of dive, but with the R1100RS there is a minimal fork-dive fed in — presumably to reassure traditional riders — but the end result is constant steering geometry whatever the road conditions or brake or throttle modes.

The natural partner for this system, which allows you to brake late and hard since you are not upsetting the geometry by banging on the brake lever, is ABS; and they marry up on this bike to make a truly impressive combination. Anyone who remembers old flat-twins will recall just how soft the front end was: ideal for touring, really rather worrying if you wanted or needed to go fast and corner hard. Now you can go as hard as you like and even brake later than on many more sporty bikes. Not bad for a buffer's bike.

(BMW (GB) Ltd.)

(BMW (GB) LTD.)

(BMW (GB) LTD.)

Within those oddly shaped cylinder heads further goodies abound. Four valves are just the start, since to fit them in required a major redesign. Exhaust valves are at the front, where they get most of the cooling air, with inlets at the rear, driven by a camshaft set to the side and back of the cylinders. The cams act on cup tappets and short pushrods to twin rocker arms operating the valves in pairs. Oil-cooling adds to the passing air to keep the temperatures down. It's complex but it works, much like the Motronic fuel injection which has finally replaced those Bings. The end result is 50% more power than from previous boxers, with a similar leap in torque to 70lbf.ft at 5500rpm — you can't help but notice a power jump that big.

The corporate parts bin got raided for the wheels, brakes and tyres, which are all off the K1, but perhaps less successful is the mating of the K-Series gearbox with the flat-twin. You still end up with a slow, notchy five gears which don't really fit in with the much more modern feel of the rest of the machine.

It certainly looks modern, which is quite a trick for a machine that fundamentally dates back to 1923. You get stylish looks, a radical front end, plenty of poke by BMW standards, and a beautifully engineered and constructed machine that tells you it has been well made every time you ride it. This is elegant engineering at its Bavarian best.

Price:	£9355
	(for fully faired SE)
Power:	90bhp at 7200rpm
Top speed:	120mph

Those who worry themselves about the overtly sexual similes that have been applied to the Ducati 916 need not be so concerned about the Monstro. The lines of the 916 do beg some comparisons, with the fairing curving lusciously over the stark metal; but the Monstro is minimalist, a brutal latticework around the finned powerplant.

The Monstro name-tag began as just a jokey working title at the factory; but it stuck, despite having less actual justification than in the case of the stunningly ugly Alfa Romeo that also bears the 'Il Monstro' tag. In its way it is a bit of a Frankenstein's monster in that it is largely the result of a trawl through the parts bin at Bologna rather than an all-new concept off the gleaming drawing board. Let's see now, we need an engine — Luigi, what are you sitting on? Aha, a 900SS unit, that'll do. What shall we hang it from? Well, we've got lots of frames and rear Boge suspension units from the 888. Right, what else is lying about? We need a front end-stick on the 41mm Showa upside-downies from the 750SS. Time for a siesta, si?

When it came time for the dials they didn't really bother since it was past lunchtime by then, so you get a speedo, idiot lights you can't read in sunshine, and absolutely nothing else. The odd thing is that the Monster is much more than the sum of its parts, or absence of them. It's a radical departure from the bum-up, head-down supersports bikes we expect from Ducati, but there are unexpected advantages to this oddity in the range. Like the ability to pull wheelies at will…

The low seat height and flat bars (although the front end doesn't actually feel as high as it looks) lets you pop man-size wheelies whenever you feel like it. You just whack it open, tug on the bars, and the key fob starts a rapid rise towards your visor. Another advantage of the keys coming up is that you don't need to use the clutch. This is a good thing, since the clutch is not the bike's strong point. Smelly clutch odours are a constant companion when trying to manoeuvre through traffic for any length of time.

This is a shame, since the Monster is ideal for weaving through traffic with its upright riding position, brilliant four-pot Brembo calipers on 320mm discs, and wide bars giving plenty of leverage. Like all other Ducatis, though, the main fun factor comes when you get out on the open road. Because of what it is, you might prefer to search out smaller roads than you would on other Dukes, since A- and B-roads are the most incredible fun on the Monster. You don't get a rev counter, but you don't want to rev the two valves per cylinder off the bike anyway. Desmo action keeps it all under control should you get hopelessly over-excited, but the 38mm Mikunis give the 90-degree V-twin buckets of torque.

You just shift when it feels right, and ride the tidal wave of torque that will help you flow round the roads in a manner that seems even faster than on the actual 888 itself. Possibly this is because of the upright riding position forcing you to hang on to those bars, so that you feel the full force of the terrific

The Ducati M900 Monster (Ducati Information Services)

*The new 600cc Ducati Monster
(Ducati Information Services)*

acceleration that is available all over the rev range so long as you keep it over about 2000rpm. Then again, at just 405lb, this is one light motorcycle, even by Ducati standards.

To be honest, the front end is rather outclassed by the rear rising rate suspension, but you have to be pressing on pretty hard before you feel it. If this becomes a problem, of course, you can simply hoist a wheelie to give the front end a rest, and then finish it off with a stoppie just in case it was getting complacent. It's the sort of bike that encourages this sort of behaviour, even in people who generally model themselves on Clark Kent *before* a trip to the phone box. This is more fun than really going for the corners since, although the handling is excellent, the ground clearance is slightly restricted thanks to those bulky silencers.

But let's not carp about detail; this is a bike that gives you truly monster amounts of fun. And if that wasn't enough, there is now the 600cc version, a yellow pocket rocket that is going to sell for £5000 but which is bound to offer just as much fun as its bigger brother. It may be smaller but, hey, we're not bothered by sexual similes, are we?

Price:	£7500
Power:	73bhp at 7250rpm
Top speed:	115mph est.

(4) Harley-Davidson FLSTF Fat Boy

There are motorcycles and there are Harley-Davidsons — a fact which invokes fury in every other bike manufacturer. How on earth do they manage to keep such brand loyalty and image, without even having to explain the occasional yawning chasm between image and reality? No doubt the Evolution engines, coming in during the mid-1980s, saved the Hog's bacon; but maybe it's more a case of that old saying — if you have to ask, then you wouldn't understand.

At around the same time as the appearance of the Evolution engines came the Softail frame, a clever device that made passers-by think the manly rider was on a rigid-tailed bike, whereas his posterior told him that he had some rear suspension tucked up there under the seat. The only downer with this set-up is that the frame needs the engine solidly mounted to it, unlike the rubber-mounted motors found in the big tourers. The result is that the 1340cc V-twin lets you know that it is throbbing away underneath you even though the foot boards are rubber-mounted. You soon learn to ride with your teeth clenched together in a sardonic smile — maybe that accounts for Arnie's expression when he rode his Fat Boy in *Terminator II*.

Things like colour scheme, wheels and bars tend to constitute the basics of model changes within the Harley range, so a Fat Boy is not dissimilar to a Nostalgia, give or take a dished wheel or two; but the image is slightly different, and on a Harley image is all. For starters, despite the Milwaukee factory's unconvincing insistence that the name comes from its porky dimensions, some historically- minded cynics have remarked that Fat Boy was the codename for the atomic bomb dropped on Japan at the end of the Second World War. (Honda have yet to name one of their models the 'Pearl Harbour').

Unlike the big Electra Glides and Ultra Classic Tour Glides, the Fat Boy stands out from the crowd because it has stripped off, discarding the acres of fibreglass fairing and panniers. This is not the Harley to take if you want to join the convoy of nostalgia freaks trekking the length of Route 66; it is more suitable for going from Chelsea to Mayfair or from Nice to Monte Carlo.

The lack of engine isolation makes itself felt on longer journeys, as do the extreme width of the handlebars and the lack of any weather protection. Instead it is better for a cruise, for a pose, when slower speeds mean that more people see you. They will also hear you, thanks to the two shotgun pipes and the glorious rumble that any 1340cc Harley makes. As ever, the engine is not about outright horsepower but more about torque from tickover up to about 3500rpm — you don't get a tachometer, and you don't really need one, since your dentist will tell you if you have been over-revving the bike.

Braking, thanks to a single front disc, is not really up to scratch, but the handling is fine if you don't push it too far — it is certainly far better than the solid rear end it apparently emulates. With 650lbs floating about, this isn't the sort of bike to try to thrash about on at licence-shredding speeds, leaving your braking to the last second and then scratching round the turn. This is more about going along for the ride, taking it steady, and swinging through medium-speed turns while that glorious rumble pours the torque out without frightening you stupid.

Even within the realms of the Harley-Davidson range the Fat Boy stands out, with its shotgun pipes and dished wheels — it remains one of the biggest sellers in the UK line-up. Changes for 1994 are confined to a revised transmission with the gear cogs having a greater contact area, cutting down on noise and vibration. The belt final drive ratios have also been lowered slightly to give marginally improved acceleration; and that is pretty much it, apart from a different starter motor gearing and some new colours. This continuity is, in a Harley, sort of comforting — a radically revised model would seem almost unethical, somehow.

Price:	£11,120
Power:	Adequate
Top speed:	
	If you need to ask...

(Harley-Davidson UK Ltd.)

Life and so

ul.

Get this close to the tank of a Harley-Davidson® motorcycle and something very important is going to change.

Your life.

At first it will be subtle – getting up earlier for the ride to work, inventing reasons to make small journeys.

Soon it will be more obvious.

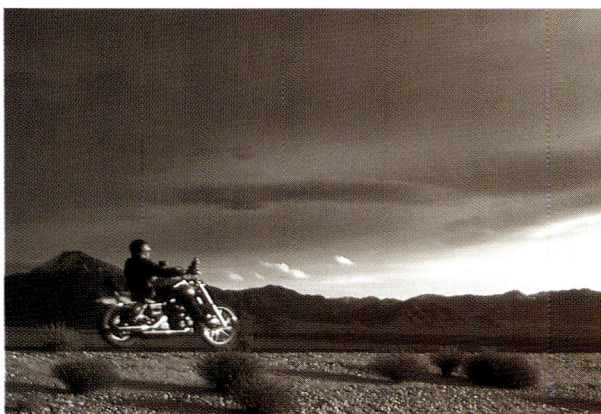

You'll find yourself taking a different way home because of a long sweeping bend you'd never really noticed before, planning weekend trips to nowhere in particular, and feeling an excitement you haven't known since you were a kid.

Before you know it, riding a Harley will be in your soul. You'll look down one day to see the tank bathed in the glow of a late summer sunset, beneath it you'll hear the lazy rumbling of a mighty Evo twin, and you'll remember when you thought it wouldn't hurt to sit on one for a moment, just to see what it felt like.

For further information on the 1994 range of Harley-Davidson motorcycles, telephone: 0280 700 101.   Quoting Ref: CAR1.

Jet fighter planes are deliberately built these days to be unstable in a straight line. Only on-board computers make them controllable under normal conditions; but during a dog-fight the lack of stability makes them perfect for rapid manoeuvering and extremely high jinks. The FireBlade feels pretty much the same.

Honda call the design concept for this bike Total Control, which means that they have taken a fresh and welcome look at how to make a bike go fast. The ZZ-R1100 described above is incredible, but it is also heavy and steady, epitomising the way most companies approach the goal of high top speed. Total Control means that the key is balance and lightness, so instead of adding more horsepower to make it go quicker they have taken off kilos. The effect is quite extraordinary, making the FireBlade a singular experience that isn't going to be to everyone's taste.

In Britain we have a notional 125bhp limit for imported bikes; some companies — like Suzuki — ignore it, while others — like Triumph — feel that it doesn't apply to them. The Honda comes in at around 120bhp and is faster than something 15bhp more powerful in most real-world situations. The engine is certainly part of this: a short-stroke DOHC four that revs like a maniac up to the 11,000rpm limit, but which will also pull incredibly strongly from low and mid revs. You get power from all over the band, building to a power frenzy that keeps your concentration like few other bikes.

Maybe something like Suzuki's GSX-R1100 is going to make more outright power, but the key to the equation is the fact that the FireBlade is 100lbs lighter than the Suzuki. That's the equivalent of taking a pillion rider off. The FireBlade consequently gives you one of the most exciting rides you are ever likely to have, feeling more like a particularly well-sorted race bike than a road machine. However, this is a double-edged weapon.

The wheelbase is pretty short, with a steep steering angle and a riding position that puts you out over the front end, splayed by the high, wide tank. Wheelies happen all the time whether you mean it or not, such is the power-to-weight ratio combined with the short wheelbase. An experienced rider is going to have a seriously good time on a bike that seems to respond to his every command instantly, without any physical input from the rider: think it, and it's happened. That means that you had better have the right thoughts. Get it wrong, and the FireBlade becomes an unguided missile.

Everyone who has ridden one comments that one of the first impressions is that it steers so incredibly quickly. For the first few corners you find yourself aiming to clip the apex, but instead you have to wobble around as you suddenly find yourself just about to apex 20 yards too early. There is none of this thinking it through, getting out of the saddle and hauling it down with a steady throttle. You just hammer up to the turn, whack on the brakes, and suddenly you're right at the apex accelerating away again.

It is so extreme that you had better pay attention. There is plenty of chat about how bikes are amazingly fast and powerful, and they are — but most are also very user-friendly. The FireBlade doesn't really fall into that category; it is too racer-biased to be anything other than frenetic. It's the sort of bike you'd want to go out for a blast on when the world was getting too much, so you could revel in the instant response, the fluid and massive drive from the engine, and the general quality of build. The rest of the time it's a nervy, twitchy rocket on which you can find yourself getting into all sorts of trouble faster than you want. Get it right, though, and it's the most perfect sports bike on the market — so long as you manage to keep Total Control.

Price:	£8195
Power:	120bhp at 10,500rpm
Top speed:	160mph

INTENSITY
ELATION
POWER

HONDA

FEEL THE HEAT

Imagine riding a bike with the power of a 1000cc machine but the proportions and agility of a Supercup 600 and you have the CBR900RR Fireblade.

The effect of its crisp power delivery and sharp handling have meant the Fireblade's been described as everything from, "Ferociously addictive," to, "The most outrageously performing production bike of all time." Bike magazine came away after a test ride and proclaimed, "No other production machine is capable of the same adrenalin rush."

We tend to agree with these sentiments, but this is what they said about the '93 Fireblade before we made some new improvements for '94 - now it's even lighter, more potent, and aggressive looking. So imagine what that's like.

For more information phone **0602 791716 (24hrs)** for a brochure and a list of Honda dealers in your area. All Honda motorcycles over 400cc carry a two year unlimited mileage warranty.

RACING WITH
Castrol

HONDA   THE SPIRIT • THE TECHNOLOGY • THE MOTORCYCLES

(6) Kawasaki ZZ-R1100

Back in 1985 Kawasaki held a press conference prior to the launch of the 1000RX. At the meeting the factory said publicly that the RX would be the fastest, most powerful Kawasaki they would ever build. At the end of 1992 Kawasaki had another launch, this time for the revised ZZ-R1100. At the press conference senior factory personnel were so concerned about the figures that they wouldn't answer questions about top speed and power. It would be irresponsible, if it wasn't such a superb motorbike

The main point of the ZZ-R is that it will do 170mph, a figure that doesn't make much sense in this country, although it does mean that owners can accelerate up to any figure they can actually handle in the absolute minimum of elapsed time. And they can make ludicrous, trouser-filling claims while standing at the bar.

Doing 100mph more than the motorway speed limit takes some doing, and an unwary or inexperienced rider is liable to find out the hard way how far ahead he or she needs to look and just how much distance reaction time covers. But at least the bike is on your side, since it is built primarily to handle that colossal speed, although the really amazing thing is that it can also be pottered round the shops without any effort at all.

There isn't any one incredible thing about the bike, it is simply a brilliant design all over. Perhaps the one area that sets it apart sounds boring, but it is the air intake that is the really clever bit. On the original version this was a sort of letter box in the fairing, but the new model has twin ram air intakes that work better and better the faster you go. Unlike many bikes, the ZZ-R1100 won't produce its best performance on the dyno unless you link up a whacking great big fan in front of it to reproduce the effect of forcing high-speed air into the intake maws.

Behind the fairing, which has obviously spent a lot of time in the wind tunnel, sits the ubiquitous liquid-cooled four, now putting out around 147bhp, although this is restricted to 125bhp for the British market. The current model runs slightly smoother at lower revs than the version it replaces, so now you have truly, madly, deeply awesome power from tickover up to 11,500rpm.

You don't need the revs since the torque is there by the bucketfull — only halfway to the redline, the ZZ-R is already making as much torque as a 1340cc Harley. And you haven't even got into the full power band. How many more facts do you need? Okay, one more — this bike will break the 70mph speed limit in first gear.

The frame is made of pressed-sheet aluminium with beefed-up steering head and swingarm pivots, so you need never worry that the bike is going to get out of shape. Suspension is to match, although the ride can be dialled in for a comfortable time even though the world is going crazy the other side of that wide and bulky fairing.

At 233kg this is not a light bike, but it carries its weight low down where you don't notice it as much as on some lighter bikes; and it does give superb stability at any speeds. This bike has won a roomfull of awards, and deservedly so. That weight and power make it expensive to run on tyres, chains, brake pads and even cush drives; but if you can stand the price you get what is probably the finest, fastest, most user-friendly bike in the world.

Price: £8350
Power: 125bhp at 9500rpm
(restricted)
Top speed: 170mph

(Kawasaki Information Service)

Get Dirty.

To fully appreciate the Cagiva Dirt Range there is nothing quite like swinging a leg over, feeling the subtle suspension, grabbing the bars and starting one of these versatile V twins or singles.

If a touring trip is planned then a choice of 900 or 750 will give you the power and comfort you need.

As your touring takes you off the beaten track these bikes will give you access to places a tourer couldn't contemplate.

These bikes are no slouches on the open road or country lanes either. Easy handling, good power and brakes can embarrass pure sports bikes.

The Singles give tremendous fun 'green laning', or on the road.

Take a closer look at our range and that trip to Morroco or a grubby Sunday afternoon need not be so far away. Visit your dealer and hear him talk dirty!

Elefant 900 from £5999

Elefant 750 £5349

W12 350 £3499

W16 600 £3799

CAGIVA
As Much Fun as you can Handle

All prices quoted do not include Road Tax, P.D.I., Delivery, Labour content of 1st service - Total £250.00 irc. VAT.

(7) Moto Guzzi Daytona

The Moto Guzzi Daytona with race kit fitted.
(Three Cross Motor Cycles Ltd.)

Moto Guzzi's heritage is almost as long as its wheelbase. The company has been making bikes remarkably like the current models for 40 years or more, giving a continuity and identity that a sector of the buying public respond to with almost religious fervour. They're the Harley-Davidsons of Italy in more ways than one. Much like the Hogs, you get a V-twin range that evolves with almost glacial speed; but with the Italian company you also get a stylish mystique that you will find only in a hot Latin country — whatever mystique might mean.

The Daytona is largely the result of work done outside the factory by Dr.John Witner and his incredible racing Guzzi, which was such a sensation at the racetrack the new model is named after. The two main legacies of this work are the four-valve heads and the parallelogram rear suspension, both there to make the bike produce more power and get it to the ground without upsetting chassis and rider. The result isn't going to worry a FireBlade owner, but the Guzzisti are going to like it.

Although four valves per cylinder often mean top end poke at the expense of lower down grunt, this has not happened on the Daytona. It still makes that sort of steam train chuffing at low revs, but the performance accompanying the noise is more like the TGV. This thing really shifts, particularly out of turns, where the incredibly generous grunt hauls you towards the next corner with surprising speed. Although there is roughly the same amount of horsepower in the air-cooled cylinders as a good Japanese 600cc bike, the 91bhp from the Daytona is spread good and thick all over the band.

Where the older Guzzis only revved steadily, with a great deal of throttle twisting, the Daytona spins up much faster and seems to be breathing much more freely. The Weber Marelli fuel injection helps, as does the stainless steel exhaust, so much so that you might actually need the rev limiter that comes in at 8000rpm. With so much grunt from 3000rpm, hitting the redline is not exactly essential, although it is true that keeping the revs up does give you a slightly faster ride.

The handling, despite the new rear end, remains very much in the style of Moto Guzzi. The 58.3in. wheelbase keeps it extremely stable without any of the twitchiness you get from shorter, faster steering bikes, but the pay-off is that you need to work it a little if you want to flick it around a tight series of bends. The rising-rate rear end almost eliminates the shaft-drive reaction you get when sawing the power on and off, but it's still not an inspired idea to change your mind even once when committed to a corner.

It's pretty comfortable, with that easy-going V-twin sensation that means you can go and go, completely locked into the bike. Not that this is a sensation you can share since there only the single seat, and long-distance riding had better be accompanied by little more than a toothbrush and a credit card; but we live in selfish times.

The rest of the Daytona has improved finish over the bikes of old, although some detailing might seem odd to someone more used to Japanese or German machinery. This isn't really the sort of bike you use day to day, whatever the weather. It isn't sensible, but it is individual, different and, of course, fast. It is, basically, something for the weekend.

Price:	£8549
Power:	91bhp at 8000rpm
Top speed:	140mph

BEAR Muscle

The Daytona has established a firm following since its launch last year. Now a major contender in the superbly popular "B E A R S" series. 3X and Raceco will again be campaigning their Daytona throughout the season.

You can see why they chose the Daytona - the tremendous mid-range rush, all-round handling and long-legged reliability.

For '94 the price at £8550.00, ready to ride away, you can experience the great performance of the Moto Guzzi Daytona road bike. Many satisfied owners have remarked on the bikes' abilities as a fast, long distance sports bike. Ring your nearest dealer and find out for yourself how much muscle the Daytona really has.

MOTO GUZZI
SUPER SPORT

(8) Suzuki RF600R

(Heron Suzuki plc)

The middleweights didn't used to get much attention, as all the effort went into the giants topping the range, but now the smaller boys are becoming the giant-killers. The 600s were really turned on their heads by the arrival of the CBR600 from Honda, but in the last year or so the others have been fighting back — notably Yamaha with the arrival of the razor-sharp FZR600R. Kawasaki used to have the most sports-tourer oriented machine with the ZZ-R600, but the new model gives up some of the overall balance in favour of storming top end stomp. That left a gap into which the new RF600 slid last year like a piston into the bore.

For a 600cc bike it seems large, and surprisingly comfortable. For a 600cc bike it has surprising amounts of low and mid-range power. You can't help but be impressed by the way these machines have improved all round in the last few years. But perhaps the most impressive thing about the RF is the engine. This is all new, not based on GSX bits and pieces. It is a very short-stroke four-cylinder (65x45mm), revving to an indicated 14,000rpm. Thanks to steep valve angles, 55-degree down-draught carbs and a whole host of tweaks there is a claimed 100bhp available at 12,000rpm — although the real figure is slightly less than this — which is pretty incredible for a 600.

What is even more incredible is that all the horses are not crowded up into the very top of the rev range. Look at the tacho and you might think you will be riding a screamer, but there is a good shove of power from low revs with just a small dip at 5500rpm and

(Heron Suzuki plc)

8000rpm. Keeping t cool is a truly massive radiator, which cools the liquid round the head and barrels as well as the oil thanks to a heat exchanger. Some retread returning to biking after a few years absence just wouldn't believe that this is only a 600cc bike, so hugely powerful is it from all over the rev range.

The six-speed transmission and clutch help this power flow smoothly and seamlessly, without any glitches. Vibration, which used to be something of a minor problem on the older models, has been almost completely eliminated. All this gives you a powerplant that you can use for balls-out riding, commuting or just touring without any great sacrifice in any department. The rest of the bike falls in with the engine.

The massive beam frame looks the business, even though it is actually made of steel, like the swingarm, unlike the aluminium efforts coming from other factories. There doesn't seem to be any great weight penalty paid for this, and there is no doubt that the frame is plenty stiff enough for anything even the most crazed rider could come up with. Suspension is set up with the sports-tourer more in mind than the racer, but it can cope with most situations, while the four-piston brakes are absolutely faultless and reassuring.

The result is a bike that you can ride across several counties without needing your personal osteopath in close attendance.

The styling is one of the bike's strongest points; not only does it look dramatically different, it is also aerodynamic, so that the rider gets a fair amount of wind protection without having to adopt a riding position that puts his helmet lower than his boots. The seat height is pretty low; this is an attractive bike that should appeal to both men and women who want something sporty but who also want something more real-world for most of the time.

Some bikes are now on another planet, with performance beyond the rider in most instances, but the 135mph of the Suzuki RF600R is still more than enough for most people — particularly those who have scored rather too many points on their licence. While this may be 35mph less than that ultimately attainable elsewhere, you are still looking at a fast, rapidly accelerating bike that won't wreck the tyres or transmission so fast and which is almost reasonable to insure. And you can take it off for a touring holiday with a pillion without taking up yoga. The 600s never used to be like this.

Price:	£5849
Power:	92bhp at 12,000rpm
Top speed:	135mph

(9) Triumph Speed Triple

In a world awash with high-tech sportsters from Japan, looking more and more similar as everything disappears behind streamlined fairings, there have been two companies that stand out from the crowd. Neither is from the East: Ducati in Italy and Triumph in Britain. Both companies have a heritage that reassures the buying public, but neither now makes a product that does anything more than hint at past glories.

Triumph lived and died and finally lived again, its reincarnation due to industrialist John Bloor who oversaw the launch of a modular range of bikes in August 1991. We've had triples and fours in various guises and strokes; but now the range is maturing, and actually offers a model with a new name. There was a Speed Twin; but there was never a Speed Triple.

The 12-valve liquid-cooled triple is now the power unit in four models — Super III, Speed Triple, Sprint and Tiger — covering everything from the 114bhp Super III sportster to the 84bhp Tiger trailie. It's a gorgeous unit, and fits perfectly into the 97bhp Speed Triple, which is the perfect epitome of the modern cafe racer. Gone is the image of big twins, open-face helmets and tea and chips; now we have a smoothie that would be equally at home at the cafe or the three-star restaurant. But in the Sixties they didn't have civilised horsepower like this.

Torque and horsepower are all over the range, but there is a further shove at 6000rpm with maximum torque of 61lbf.ft at 6500rpm. Top speed is around 135mph, although that is really only attainable by those with neck muscles like Mike Tyson. The engines are well put together in the UK, using castings produced domestically although things like the crank and camshafts are made in Germany and come to Hinckley to be finished off. The result is smooth and gorgeous, in a way that only well-put-together triples can be. This is a long way from the 180-degree Laverda Jota which could numb you worse than a nighshift on the road drills.

Power is flowed through a five-speed box, which is almost old-fashioned but which will come as a relief to those who have had to suffer the fashion of six-speeders even though the engines patently produced the power and torque to cope with five. Being chain-drive, you could always fiddle with the rear sprocket if you felt like going for a more relaxed ride at the expense of some acceleration.

And of course there is one further benefit to riding a triple — the aural dimension. At tickover it must be said that some Triumph triples sound like diesels, but once into the midrange those two pipes with their carbon-fibre lookalike covering start to emit some really serious stimulation. Just crack it open, feel the shove and listen to the band.

That power unit is really the best of it, although the rest of it fails to let it down. The handling is courtesy of the spine frame which is built in-house, riding on 43mm front forks with triple-rate springs and Kayaba rear shock (which is only adjustable for rebound damping). The riding position puts you comfortably over the bike, without the feeling that somehow you've sat on the handlebars — FireBlade riders must know the sensation. Steering is steady enough and is stable and reassuring rather than linked directly to the brain. The result is an ability to drive hard through the turns without any jittery worries about apexing too early or terminally upsetting the whole plot. Those who like to wear out their knee-sliders might find the pegs decking too early; but really, if you are going for the absolute max then get the Super III — if you can afford it.

While the Super III looks terrific, particularly in yellow, the Speed Triple, which is mostly sold in deepest black, looks even more carved from solid. The white dials (red at night) look just right, and the alloy fascia continues the quality retro feel without sacrificing any of the recent advances. That's the point, really: you get Nineties biking with just enough of a hint of the past to satisfy young and old. People like the past, but they'd prefer Nissin four-piston calipers and Michelin Hi-Sports to drum brakes and TT100s. The Speed Triple shows that this is going to be a good time to get nostalgic about in the future.

Price: £7499
Power: 97bhp at 8000rpm
Top speed: 135mph

(10) Yamaha Diversion

The Yamaha range has always had bikes which seem to hang in there well past their best-by date, like the XJ600 and the FJ1200. Both are mighty fine products, eminently usable on our busy roads and with origination costs that were amortised years ago, leaving riders not bothered by the razor's edge of fashion to enjoy biking at a reasonable price.

Then a couple of years ago Yamaha revamped the XJ and called it the Diversion, although in reality — and even in marketing terms — what we have here is an FJ600, half of the mighty air-cooled four that has sold well in one form or another for over ten years. Middleweights make a lot of sense, particularly if you stay clear of the repli-racers, and particularly if you care about insurance, fuel bills, tyre and chain wear, and even the state of your licence. Whereas the FJ12 is going to gulp a gallon down every 35 miles or so, the Diversion is going to give you over 60mpg if you play it cool, although you will be getting 60bhp for

(Mitsui Machinery Sales (UK) Ltd.)

(Mitsui Machinery Sales (UK) Ltd.)

your fuel instead of 125bhp. In this day and age that is not a whole lot of power, compared with, say, the Suzuki RF600R (below) which makes over 50% more horsepower than the Diversion. But to get that power Suzuki has come up with an all-new, liquid-cooled engine that uses every trick in its book to squeeze out power. The Diversion hasn't read that particular book.

Instead you get an air-cooled four, canted down to help cooling and a good inlet charge, a tubular steel frame and only one front disc. The clean lines help disguise what is basically an ancient design. The funny thing is, it still works rather well. Although it all looks like stuff taken from the parts bin that was sealed for posterity, the bike is actually all new from the ground up, so some tweaks have worked their way into the Diversion.

Although you only get 60bhp at 7500rpm, this means that the power is spread fairly evenly over the band, not squeezed up into the five-figure part of the rev counter. Coupled with a six-speed gearbox — five would have been more in keeping — you can keep it churning over nicely up to about 100mph, at which point it is starting to lose interest in doing anything dramatic. Another by-product of the engine is that the buzzy vibration that you got with the XJ and other middleweight fours seems to have been pushed more into the background.

The steel frame holds things together fine, while normal telescopic forks at the front and a monoshock rear end keep things in line even under the sort of riding that pushes the engine to the limit. But this is not the sort of thrashing that most Diversions are often going to get, deserved or not. With that in mind, there is a goodly amount of protection from the half fairing,

while the riding position is a good compromise for riders of even remotely normal dimensions. The pillion certainly gets a fair old chair, far more comfortable and spacious that those found on bikes of double the capacity.

This all combines to give you a bike you could happily use for commuting, and which wouldn't look out of place in an executive car park. The dark green scheme in particular looks very exec. But then you could take off for the weekend, and the Diversion stands out among most other middleweights in that it offers a serious touring option for both rider and pillion. In a world of Gatsos, 200 tyres and roadworks, the Diversion is one route well worth exploring.

Price:	£4399
Power:	60bhp at 7500rpm
Top speed:	120mph

The Classic Bike Marketplace

The "classic" motorcycle marketplace is a minefield for the unwary and the over-optimistic. Indeed, although premium prices are obtainable, it is only the expert and specialised trader who is able to manage this — at least consistently.

Prices in the classic car world suffered from a catastrophic melt-down in the opening year of this decade, with considerable sums being lost as the trading values of old motors fell floorwards. And, as ever, the motorbike market followed suit, with the prices obtainable for top-end classic bikes being severely weakened for similar reasons.

In the closing years of the Eighties, with prices for older — or "classic" — vehicles rising on the general tide of optimism which characterised the Lawson Boom, it began to appear that the public's demand for old vehicles (and for rare stamps, furniture, boats, and every other type of "durable") was insatiable. Sterling was reasonably strong against the US dollar, and it made increasing commercial sense to dig out a secondhand stack of those vehicles which had been a mainstay of British export effort through the "classic" years from the end of the last war until the early Seventies.

And so it happened, with container-loads of repatriated British iron — both four- and two-wheeled — being floated back across the Atlantic. The condition of these vehicles was generally rather better than found on those of a similar age which had endured the salty depredations of several British winters. Indeed, many models were export types of motorcycle often only familiar to British enthusiasts through the pages of books and American magazines. Since these bikes were also frequently quite cheap, they sold well.

Sadly, the turn of the new decade saw many changes, not least to the amount of money available to the two-wheeled enthusiast, who is generally less affluent than his four-wheeled counterpart anyway, at a time when the classic bike market was flooded with

large numbers of slightly tatty (and therefore fairly cheap) ex-American market British bikes. The inevitable price-softening took place as market forces swung inexorably into motion, and the old bike market drifted — in company, it must be said, with the new bike market.

The dying months of 1992 were perhaps the quietest of all. Early 1993 saw a gradual improvement in trade, and this year so far appears to be looking better again. Although the number of bikes being moved does not appear to have grown remarkably, the machines are selling, and at sensible prices. The figures which can reasonably be asked for Vincent and Velocette machines, for example, are nowhere near the elevated levels of a few years ago; but they are at least rising again after some serious falls.

The classic bikes themselves fall into three reasonably definable categories. First, and least numerous, are the purely collectable machines, which demand premium prices because of

Above: Triumph twin-cylinder motorcycles are the mainstay of the classic bike market, with more unusual examples, like this 750cc TSS commanding high prices.

Left: A large number of previously exported machines have been re-imported from around the world in recent years. This Triumph 750cc T160 Trident was brought back from the USA in 1992.

Right: *While there will always be a market for the older — ie. pre-1965 — bike, workaday machines like this 500cc Matchless from the mid-50s are rather less popular than later twin-cylinder models.*

Below: *Although Norton is one of the most charismatic marques from the British classic period, their small capacity models — like this 350cc Navigator — are not generally in high demand.*

their perceived rarity and desirability rather than because they are pleasant to ride. Bikes in this category include Vincent twins (but not the singles which, although worthy enough, lack the 1000cc twins' charisma and fire-breathing reputation); sporting Velocettes (the Venom and Thruxton are the two most in demand); and camshaft Norton singles (International roadsters and Manx racers always collect high prices, and are most popular with overseas customers).

Not all of the machines which have undoubted rarity are desirable or command high prices, however. There are several models, from many manufacturers, which are rare today either because nobody bought them when they were new (because they were horrible), or because they were all scrapped not long after purchase (because they were unreliable as well as horrible). A good — if that's the word — example of a famous manufacturer themselves in the foot was Norton's Model 77. This unlovely, unloved and uncommon motorcycle combined the 600cc twin engine from the fine-steering Dominator 99 with the bendy chassis of the ES2 to produce a machine supposedly intended for "the sporty sidecarist", whoever she might be. To the punter of the mid-Fifties it looked like an exercise in using up over-stocked ES2 chassis, and it hasn't matured gracefully.

Of course, there are Norton "completists" who would no doubt hail the Model 77 as a pinnacle of something or other, but there cannot be many of them.

The second class of classics, which are the most numerous and probably the most popular, are the fashionable 650 and 750cc twins from Triumph and the 750 and 850cc twins from Norton. Both marques built their twins in large quantities and in several versions, and huge numbers of both have been repatriated from the UK's major export markets of North America and — to a rather lesser extent — from South Africa and some other African states (Nigeria was once a major customer for military-spec Triumph 750cc twins, and several container-loads have been re-imported). In addition to Triumph and Norton, BSA and to a smaller degree AJS, Matchless and Royal Enfield

models are always popular enough, although they are generally less sought-after and tend to attract lower prices.

Finally, there are an awful lot of machines which were unearthed and hyped as "classics" during the selling frenzy of the late Eighties, but which may charitably be described as undesirable. These are the grey porridge machines of the Fifties: machines which offered slow and faintly unreliable commuter motorcycling when they were new, and which were thrown into the backs of sheds (or canals) when they failed to start for the last time. These bikes were built down to a very low price when they were current; they were pretty uninspired then, and they have most certainly not improved with age. Although there is always likely to be a buyer somewhere for absolutely anything, it would be a brave soul indeed who would fill a showroom with Francis- Barnetts, James scooters, and the products of all the other Villiers-powered assemblers who were driven to the wall when Honda unsportingly offered novelties like reliability and comfort with the introduction of the C50.

Left: *One area of the classic market which is showing signs of increasing interest is that of the European classic. Large capacity Italian machines, like this Ducati Darmah 900 are in particular demand.*

Below: *Customised machines, like this Norton Atlas in 'café racer' trim, tend to sell less well than similar models in standard, factory condition.*

It is of course impossible to categorise so glibly all of the myriad machines available; and all of the popular manufacturers built horrors as well as heroes.

In addition to the three British groupings offered above, there is a steady growth of interest in older machines from Japan and from continental Europe — more specifically from Italy.

The accelerating interest in "classic" motorcycles from Japan was inevitable, as riders who cut their teeth on the products of the Seventies started to return to motorcycling after the familiar mating and breeding hiatus. Indeed, the Japanese themselves — who are major collectors of old British machinery — recognised this trend when they introduced the current range of retro-styled modern machines targeted directly at the returnee biker. Whether a new Kawasaki Zephyr manages to offer quite the same character as a genuine Z1 is a moot point; the low-mileage returnee is often satisfied only by acquiring an example of the sort of bike he owned (or aspired to) before he lost his youth.

And once again, the market has been expanded by the import

of large numbers of low-mileage, good-condition motorcycles from the USA. Similarly, this influx of unusual (in the UK at least) models has kept interest in the bikes themselves high while also keeping the prices low.

There are many oriental oddities, too, and the classic classifieds regularly feature bikes like Suzuki's GT750 2-stroke 3-cylinder "radical kettle" and the same manufacturer's RE5, the Wankel-engined heavyweight which sold as well as it looked when it was new, and which is being hyped as "unique" by committed optimists in the ads.

Large Italian 4-stroke motorcycles have the happy distinction of being hailed as classics while they are still in the crate, and in many cases the title is justified. Certainly, there is no shortage of riders who would agree that the big twins from Moto Guzzi, Ducati and Laverda are both good and desirable. Many others would aspire to Laverda triple ownership if they had the courage, and MV Four ownership if they had the money. A couple of classic dealers of my acquaintance have confirmed that large capacity Italian machines, of any age, move from their showrooms as well as do the ancient Brits. The same cannot, apparently, be said of elderly BMWs.

Interest in Italian bikes has once again spurred the import of a fair number of usually small capacity machines from big name manufacturers. Although a 125cc single from, for example, MV Agusta may be of interest to collectors, it is of little use to anyone who wishes to actually ride their classic on the roads of the Nineties, and they do not sell very rapidly.

Pre-war machinery is also readily available, but the fact that their prices have remained static for years suggests that the market is not growing and that the supply of machines is correct for that size of audience. The last decent price hike in pre-war motorcycles has probably got more to do with the trade in cherished registration numbers than with a sudden growth of rider interest in the motorcycles of fifty years ago.

The key to satisfied and successful negotiation of the classic bike market — either as buyer or seller — lies in a decent level

Right: *The most popular machines are always going to be those with sporting pretensions. The early Triumph Bonneville and a great rival of its time, the AJS 31CSR — pictured here — will always command premium prices in an increasingly knowledgeable marketplace.*

of knowledge. It is depressingly easy to end up owning a machine which is either quite unsuited for its intended purpose, or which is not even what it was claimed to be by its last vendor. In fact, herein lie the two major causes of dissatisfaction among classic motorcyclists, be they born-agains or style refugees who wish to be seen aboard something bright, shiny and loud which is neither new nor Nipponese.

Firstly, if a classic pilot wishes to own a motorcycle which he can use everyday on the modern highways, then he or she should be sold either of the two most popular twins: Triumph's 750cc Bonneville or Norton's 750 or 850cc Commando. Spares for both of these models are readily available and easily affordable; both bikes look butch as hell and sound better than that; and really sound examples are readily available retail for under three grand. Selling an early Sixties Norton 250cc twin, for example, with its dim 6 volt electrics, pathetic performance, weedy front brake and bizarre styling to someone who wishes to be an active rider on a bike which looks and sounds distinctive but which is quick and reliable, would not be an act of charity — or of good sense, because it would return to the showroom faster than a bounced cheque.

Likewise, with the growth of interest in older bikes came a similar interest in unusual variations on common themes — usually rare sporting versions of models produced for the US market: Triumph TT Bonnevilles and BSA Gold Star Catalinas are good examples of these, as are Norton and Matchless 750cc desert sleds. Hand in hand with this trend came the artful bodger, who converts a more common, and accordingly less valuable bike into a very rare and desirable oddity. So if you are offered a BSA Rocket Gold Star, for example, take along an expert competent to ascertain that the bike is exactly what it is claimed to be, and not a common-or-garden Golden Flash with dropped bars and chromed mudguards!

The classic scene is alive and well. Prices are starting to rise again as more cash becomes available in the economy and in the pockets of enthusiasts. There are profits to be made, but there are pitfalls galore. Buy — and sell — with care.

Frank Westworth
Editor
Classic Bike Guide

In the Right Gear:

Clothing & Accessories

Look, we're reasonable people. We think you need some protection, and dat's gonna cost ya. How much? Hey buddy, get real — you know it ain't dat simple. The more you pay the less chance of somethin' nasty happening; so, lemme ask you how much you wanna pay?

It's all about protection money. Protection starts with the primary safety of knowing what you are doing, so you need to be trained. Companies like CSM, with branches nationwide, will make sure that you stand less chance of ending up on a truck grille, a situation that is going to be terminally horrible no matter how smart your new leathers and dark visor. The protection extends to keeping your bike in good condition, with good tyres, well adjusted chain, sound brakes and nothing coming loose — it most definitely extends to having two mirrors that work and which you look in every few seconds.

If the situation deteriorates rapidly and unpleasantly you are going to need a helmet that not only has trick graphics, but which also protects your head and which will not come off in an impact. When you try one before buying it, make sure it is tight, and do up the strap. Then get your hands under the back lip and try to roll the helmet up and off your head. If it moves much, try something else. Each country, and even some companies within that country, make helmets for different basic head shapes: so a German helmet may well not suit your sleek little otter's skull as well as a British helmet like an Everoak, or a Japanese one like the excellent lids from Shoei or Arai. Italian helmets, which are often stylish and less expensive than oriental offerings, can make a good compromise — look at AGV, Nolan or MDS.

Helmets are covered by standards, like the British Standard 6658A or B which your helmet should carry; but a product that can cost five times the £300 of an Arai Quantum will carry no kite mark to help you make up your mind. Leathers are chosen as much for fashion as for function, but one day the decision on which set you choose may make the difference between a good story down the pub, and months in hospital having skin grafts. Of course, not everyone can afford leathers anyway; but at the very least you owe it your tender little body to get a leather jacket and preferably leather jeans, making sure you get cowhide and not sheepskin. Leather gloves are an absolute basic requirement, and leather boots come almost as high on the list.

Leathers mostly come in cowhide, although some are now made from more exotic creatures like goat or horse. You need to be your own quality controller, so don't be blinded by the fact that they look pretty and have armour in the elbows. Check that there is double stitching on all the main areas where patches are brought together, like the shoulder, hips, knees and so on. Ensure that the leather feels reasonably thick — 1.2mm is an okay minimum — and looks the same quality all over. Some panels may have a stretched, wrinkled look to them: a suit like that should be avoided, since this is the sign of axilla skin taken from the armpit of the cow, which is not as strong as the rest of the beast.

Above: The Alpina jacket from the Gear range of leathers by Euro Helmets Ltd. of Kirkby-in-Ashfield, Nottingham; made from 1.2-1.4mm cowhide, this has Kevlar Keprotec reinforcement at shoulders and elbows.

Below: The Max jeans from the Gear range, with Keprotec reinforcements at hips, knees and calves and zip attachment for their Max or Alpina jackets.

Above: Cruise salopettes from the Gear range, with full height panels front and back.

In a good suit lots of panels means lots of pieces of leather stitched onto a basic suit underneath; but on a cheap suit the same design means lots of little bits of leather all going together like a jigsaw. "Leather is strong, seams are weak" is a useful maxim, so choose a suit with the least number of dodgy panels — if you feel behind a panel you should be able to feel a second piece of leather behind the first. That is the sign of a good suit — as long as it is made of decent leather with careful, double stitching.

Companies like Ducati, Harley-Davidson, Bimota and Yamaha make their own leathers and accessories — or get them made for them, to be more accurate — but BMW probably comes top of the list with its expensive and very fine leathers made for the sort of people who buy the bikes. You won't find garish goatskin but you will find quality; and of course you could top it off with the company's own helmet, boots and gloves, as well as a host of outdoorsy-type stuff from its Active Line range.

Nankai, Kushitani and Dainese, among others, get maximum exposure on the grand prix circuit by clothing the top stars; but they pay good money for that, and Kevin Schwantz's Dainese leathers are not going to bear much resemblance to what you buy off the peg. British companies like Fieldsheer and Frank Thomas are working their way in, however, and are making better and better leathers for the road or track, in competition to the German company Hein Gericke, which is making an increasing impact over in Blighty with its range of gear.

Right: Power Aquapak oversuit from the Frank Thomas range.

Above: *Frank Thomas range Race Sport blouson jacket incorporating Kevlar and Armasport protection.*

Below: *Defender jacket and jeans from the range of Frank Thomas leathers, with Armasport reinforcement.*

They will offer you a selection of leathers — as will Euro Helmets, who are typical in that they offer a basic leathers range made abroad (the Rhino range), and a better made and better specified selection (Gear leathers) designed and built in the UK. You can get anything to suit your pocket; but go for a well-made set, even if it looks a little duller than the day-glo one-piece hanging up next door to it.

The British company Frank Thomas also make leathers in a striking array of colourways, like many others incorporating kevlar and body armour in most suits at strategic sites. Body armour has found its way into oversuits and rain garments, although the softer material's ability to hold armour in place while sliding down the road at speed has to be questioned. Companies like Frank Thomas and Hein Gericke offer a full range of one- and two-piece oversuits, lined and unlined. There is obviously still a place for traditional wear like that from Belstaff, and the excellent oversuits which have been worn for many years by despatch riders — Rukka and Motomod aren't particularly cheap, but ask someone who is freezing cold and wet and who still has a hundred miles to ride how much they would pay to be warm and dry.

Two-pieces are more adaptable, but a one-piece tends to fold up smaller, and is less likely to billow out the back — an amusing effect which does nothing for your drag coefficient nor the insulation of your spine. Unlined suits won't keep you warm although they should keep you dry. They will fold up small and so tend to get carried around more. Most are cut fairly generously, but when you buy one make sure you try it on over your normal riding gear — it may well fit beautifully over your sweater and jeans, but then you find you can't get it on over your leathers. About £40 will buy you a reasonable unlined one-piece, although you could easily pay three times that for a top-line suit. Lined suits obviously cost more, although you could get a Sidi suit for around £60. Then again, you could pay £600 for the top Difi suit. That much would buy you a pretty good set of leathers.

Sales of leathers have increased significantly in the last few years, partly as a result of more and more riders buying bikes as a leisure pastime and not as a commuting necessity. Black is still a popular colour — the entire Gear range, for example, comes only in black — but many riders want to mimic the grand prix stars and their mulit-coloured confections. As a result the modern male rider will go out in pinks and purples that form no part of his normal butch wardrobe. With many bikers returning to the fold after years of breeding at home and battling at work, there is a marked increase in the leisure sector money, with the result that accessories like knee-sliders are now virtually an essential item on most leathers (even if they have only been ground away with a file at home to achieve credibility).

Of course, such items do highlight the progress that this influx of professional money has brought in its wake. Tyres used to be round and black and were only changed once a year; now there is a market for tyres that will cost a couple of hundred quid, but which will only last 2,000 miles or so if the considerable performance of the latest generation bikes is used to the full. Companies like Avon, Metzeler and Michelin will make you a radial that will allow you to explore the full potential of the bike's lean angles and your own courage; but be prepared to pay the price, and not just in worn-out knee-sliders.

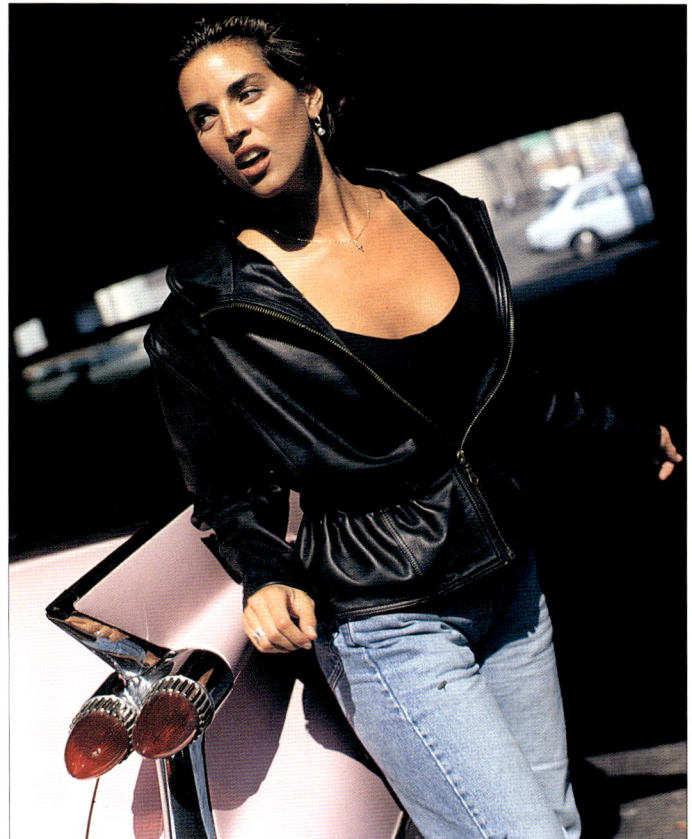

Above: Cowhide San Remo ladies' jacket from Hein Gericke range.

Left: *The Kushitami range of one-piece armour-reinforced racing suits are available from Hein Gericke.*

Below: *Hein Gericke's Columbo cowhide jacket, with quilted thermal lining.*

Similarly, the technology that we have now will enable the rider of a middleweight to get a significantly greater mileage out of his tyres thanks to the cool running of the radials. And with the increased performance of the middleweights, coupled with their big-bike styling and added attraction as insurance premiums move ever upwards, there is now a bigger and bigger marketplace in which you can delve to find the right niche for yourself and your wallet. This would seem to be the way of the future because it seems to be what we want.

The days are passing when the little dealership on the green gets your custom because, much as we all like to see him there, we tend to go off to a bigger dealer to buy our bike and fight over the discount. Bigger should and sometimes does mean more choice and more availability of those tiresome and expensive spares, as well as the whole bike service industry that goes with the market.

The Carnell group, with three big dealerships, is a typical case in that it offers you the whole lot — bike, spares, accessories, clothing, tyres, knee-sliders. Peter Bowler, a Carnell director, sees this as the way of the future. Smaller dealers either went to the wall in the recession or are going to be fighting for survival against some seriously large competition. The way ahead, according to Carnell, is to offer what the customer wants. And that is a relaxed environment, a service which instils confidence in the operation and, most importantly, a recognition that the customer is there because it is his leisure pastime, not because he has to be.

Traditionalists may dismiss this approach, but it would appear to be the shape of the biking world to come — the world of a proud rider with his helmet from Japan, his leathers from Italy, his bike from Britain, his tyres from France, and his insurance from Hell.

DIRECTORY

Section I

Bennett
The Briars, Burnby Lane,
Pocklington, Yorkshire,
YO4 2QD, UK

Tel: 0759 304753

**Bruce Main Smith &
Co. Ltd**
132 Saffron Road, Wigston,
Leicestershire, LE18 4UP,
UK

Contact: Don Mitchell
Tel: 0533 777669

Over 2,500 photocopy manuals,
spares lists and brochures and all
new motorcycle books - send for
lists. Largest stock of original
out-of-print literature.

**Clubman Racing
Accessories**
P.O.Box 59, Fairfield,
Connecticutt, CT06430,
USA

Tel: 203 256 1224

**Condor Motorcycle
Developments**
Warwick House, 59/60
Derby Square, Douglas,
Isle of Man, UK

Connoisseur Carbooks
11A Devonshire Road,
London, W4 2EU, UK

Tel: 081 742 0022
Fax: 081 742 0360

Car and motorcycle books
worldwide mail-order service
Bookshop open six days a week.
Many imported books and other
rare titles stocked.

Delta Press Ltd
Building 8, Billet Lane,
Durrants Complex,
Berkhamstead,
Hertfordshire, HP4 1DP,
UK

Tel: 0442 877794
Fax: 0442 877828

Mail order motorcycle and
automobile books and
wallcharts. We are agents for
Haynes (manuals and
motorcycling books), Lucas and
Bosch. Free catalogue and quick
service.

Going Places
BBC Radio 4,
Broadcasting House,
London, W1A 1AA, UK

J H Haynes & Co Ltd
Sparkford, Nr.Yeovil,
Somerset, BA22 7JJ, UK

Tel: 0963 40635

Haynes produce over 150 titles
for the motorcyclist, ranging
from the ever popular Owners
Workshop Manuals series to
many general motorcycling
books.

Merlin Books
P.O. Box 153, Horsham,
Sussex (W.), RH12 2YG,
UK

Tel: 0403 257626
Fax: 0430 257626

Motorcycle books, videos,
workshop manuals and more.
Free catalogue (U.K.), phone, fax,
write now. Phone until 8.30pm.

Mill House Books
The Mill House, Eastville,
Boston, Lincolnshire,
PE22 8LS, UK

Tel: 020 584 377

Rolling Thunder UK
406 Staines Road, Bedfont,
Middlesex, TW14 8BT, UK

Tel: 0629 57901

Shirlaws Motorcycles
92 Crown Street,
Aberdeen, Grampian,
AB1 2HJ, UK

Contact: Roy Shirlaw
Tel: 0224 584855

Mail order on bikes by BMW,
Ducati, Triumph, Yamaha,
Suzuki, Honda, Vespa. Parts,
clothing & helmets, specialist
repairs, training. Motorcycle
world all under one roof.

CLOTHING

Alan Dufus M/C
St Clair Street, Kirkcaldy,
Fife, UK

Tel: 0592 264135

Alvins Motorcycles
9B Springfield Street,
Edinburgh, Lothian
EH6 5EF, Scotland

Tel: 031 554 4155
Fax: 031 553 3093

Main dealers for
Harley-Davidson, Suzuki, and
Ducati; full sales, spares and
service facilities. Also Scotland's
largest and best stocked clothing
and helmet centre.

Anson Sports Products
18 Elm Street, Ellesmere
Port, South Wirral,
Cheshire, UK

Tel: 051 356 1601

Army Shop Ltd
11 Greens End, Woolwich,
London, SE18 6HX, UK

Tel: 081 317 2324

Axo Sport Dealer
Serval House, Clifton Road,
Shefford, Bedfordshire,
SG17 5AE, UK

Tel: 0462 815757

BKS Leather
27-29 New North Road,
Exmouth, Devonshire,
EX8 1RU, UK

Contact: Brian Sansom
Tel: 0395 278861
Fax: 0395 225244

Specialist manufacturers of
bespoke leather motorcycle suits.
Made to measure for racing and
touring. Suppliers to 25 British
Police forces and top
international road racers.

Bike Gear
Admail 38, Kettering,
Northamptonshire,
NN15 7HH, UK

Bikers Gear Box
144 North Parade, Matlock,
Derbyshire, DE4 3NS, UK

Tel: 0629 57901

**Bikes 'n Shirts
(Dept.WG)**
9B Market Street, Whaley
Bridge, Stockport,
Cheshire, SK12 7AA, UK

Tel: 0663 732424

Motorcycle makes/logos/
slogans on adults/kids t-shirts/
sweatshirts, badges/patches,
goggles, custom/classic/
standard bike parts. Send SAE
for catalogue. Mailing address
only.

**Bill Brown's Motorcycle
Centre**
High Street, Whitehaven,
Cumbria, UK

Tel: 0946 692697

Bon Accord Leisure Ltd
PO Box 2121, Shirley,
Solihull, Midlands (W.),
B90 4BP, UK

C.H. Biggadyke M/C
23/27 Westlodge Street,
Spalding, Lincolnshire, UK

Tel: 0775 723037

Carrick Motors
62 Queen Charlotte Street,
Edinburgh, Lothian, UK

Tel: 031 555 2575

Castle M/C
7-9 Bridge Street, Aire
Street, Castleford,
Yorkshire (W.), UK

Tel: 0977 553523

Cissbury Leathers
2 Nepcote Lane, Findon
Village, Sussex (W.),
BN14 0FE, UK

Tel: 0903 877055

Road and race specialists. Open
seven days 9.30-6.00. New and
secondhand suits. Mail order
service available.

Classic Covers
Tyn-y-Felin, Penmynydd
Road, Llangefni, Anglesey,
UK

Tel: 0248 722041

Cobb & Jagger
3-7 Saltaire Road, Shipley,
Yorkshire (W.), UK

Tel: 0274 591017

**Cosmopolitan Motors
Ltd**
220 Old Kent Road,
London, UK

D H Autos
6 Merriel Street,
Newcastle under Lyne,
Staffordshire, UK

Tel: 0782 613889

Dainese Bike Shop
104 High Street, Stevenage,
Hertfordshire, SG1 3DW,
UK

Tel: 0438 317038

Daytona
42-48 Windmill Hill, Ruislip
Manor, Middlesex,
HA4 8PT, UK

Tel: 0895 675511
Fax: 0895 630654

Dealership for Kawasaki,
Triumph and Ducati
motorcycles. Also main parts
stockist for Kawasaki.

Dorset Dirt Bikes
345 Ashley Road,
Parkstone, Poole, Dorset,
UK

Tel: 0202 721500

Dragon Skin
The Dragons Lair,
5 Catherine Hill, Frome,
Somerset, BA11 1BY, UK

Tel: 0373 452119

Drayton Croft M/C
14-20 Stockwell Head,
Hinckley, Leicestershire,
LE10 1RE, UK

Tel: 0455 637654

Essex West Road
77-79 West Road,
Westcliff-on-Sea, Essex, UK

Tel: 0702 353728

Euro-Helmets Ltd
Sidings Road, Lowmoor
Road Ind.Estate, Kirkby in
Ashfield, Nottinghamshire,
NG17 7JZ, UK

Tel: 0623 757262

Manufacturer and distributor of
AGV helmets, Gear leathers,
Rhino leathers, boots and gloves
and Eurodesign luggage.

Exilla Leisurewear
25 Ddole Industrial Estate,
Llandrindod Wells, Powys,
LD1 6DF, UK

Contact: P & D Jones
Tel: 0597 822884
Fax: 0597 822884

Your team/club members will be
proud to wear clothing bearing
your logo. Contact us for details
of our range of embroidered or
printed garments.

F.C.L. Racing
Carrick House, St.James
Place, Cranleigh, Sussex,
GU6 8PR, UK

Tel: 0483 275868

Five Ways M/C
17-19 Walton Street, Hull,
Humberside (N.), UK

Tel: 0482 55023

Frank Thomas Limited
Atlanta House, Midland
Road, Higham Ferrers,
Northamptonshire,
NN9 8DN, UK

Tel: 0933 410272

Garland & Griffiths
8 Snowdrop Lane,
Haverfordwest, Dyfed, UK

Tel: 0437 768434

Ad 4 Leathers

George Clark M/C
Peel Garage, 40 Whalley Road, Accrington, Lancashire, UK
Tel: 0254 385025

Get Shirty
Unit 10, Colhook Industrial Park, Petworth, Sussex (W.), GU28 1LP, UK
Tel: 0428 707645

Gizmo Marketing
Moorend Workshops, Highfield Rd, Idle, Bradford, Yorkshire (W.), BD10 8QH, UK
Tel: 0860 892055

Grand Prix Promotions
Brands Hatch Autostore, Brands Hatch Circuit, Fawkham, Dartford, Kent, DA3 8NG, UK
Tel: 0474 879524
Fax: 0474 879850

Manufacturers of high quality corporate, sponsors, team and team supporters clothing i.e. Team Lucky Strike Suzuki - enhance your image - use the professionals - Grand Prix Promotions.

Guy Fensome TS
Lamesley House, Howe Street, Chelmsford, Essex, CM3 1BA, UK
Tel: 0245 362473

Mail order suppliers of ex-military clothing, boots, gloves, wet weather and camping equipment. Catalogue available on request. Personal callers welcome by appointment.

Hein Gericke (UK) Ltd
35 Blossom Street, York, Yorkshire (N.), YO2 2AQ, UK
Tel: 0904 655379

Hog & Chop International
48 Manchester Street, Cleethorpes, Humberside (S.), DN35 7QG, UK
Tel: 0472 697859

Hursts Auto Complex
Boucher Road, Belfast, Co. Antrim, UK
Tel: 0232 381721

J Morgan
Brookfield Mill, Crumlin Road, Belfast, Co. Antrim, BT14 7EA, UK
Tel: 0232 757720

Jack Machin M/C
65 High Street, Lincoln, Lincolnshire, UK
Tel: 0522 512887

Jaybee M/C Ltd
96-98 St.Peters Road, Great Yarmouth, Norfolk, UK
Tel: 0493 855240

Jim Allan M/C Ltd
98 Grahams Road, Falkirk, Central, UK
Tel: 0324 20111

Just Bikers
144 Cornwall Street, Plymouth, Devonshire, PL1 1NJ, UK
Tel: 0752 222000

KME Racing
Pole Position, 1 Nene Road, Burton Latimer, Northamptonshire, NN15 5QZ, UK
Tel: 0536 725604

Kawasaki Auotorama
Bradford Road, Batley, Yorkshire (W.), Yorkshire
Tel: 0924 461112

Ken's Motorcycles Ltd
195 & 246-255 Westgate Road, Newcastle upon Tyne, Tyne & Wear, NE4 6AQ, UK
Tel: 091 232 1793

Leslie Griffiths Motors
10A Ewenny Road, Bridgend, Glamorgan, Mid-, UK
Tel: 0656 661131/2

Lloyd Cooper M/C
61 Queens Road, Watford, Hertfordshire, UK
Tel: 0923 221125

Martin's Leathers
Huntingdon Street, Nottingham, Nottinghamshire, UK
Tel: 0602 507912

MW Leathers
18 Barking Road, Imperial Mews, London, E6 3BP, UK
Contact: Mike Willis
Tel: 081 471 3933
Fax: 081 471 3933
Motorcycle leather clothing for sport and road. Standard or made to measure; mail order available. Visa or Access accepted. Ring for appointment.

M/C Accessories
160 Belgrave Gate, Leicester, Leicestershire, UK
Tel: 0533 624983

McLaines
144 Holloway Road, London, N7 8DD, UK
Tel: 071 607 8413

Midland M/C
30-40 Campbell Street, Northampton, Northamptonshire, UK
Tel: 0604 37551

Moto Elite
2A Woodstock Street, Hucknall, Nottinghamshire, NG15 7SP, UK
Tel: 0602 637728

Motorcycle City
336 Battersea Park Road, London, SW11, UK
Tel: 071 924 5794

Motorcycle City
533 Staines Road, Bedfont, Middlesex, UK
Tel: 081 751 2170

Motorcycle City
55 Forton Road, Gosport, Hampshire, UK
Tel: 0705 581226

Motorcycle City
470 Oxford Road, Reading, Berkshire, UK

Tel: 0734 596962

Nevis Marketing Ltd
Units 8-9 Wilverley Road, Christchurch, Dorset, BH23 3RU, UK

Tel: 0202 499533
Fax: 0202 499343

Exclusive UK distributors of Baleno waterproofs, Held luggage, IXS leathers, Spyball alarms, Falco boots and Grex helmets.

Powerline Limited
Marion Place, Port Grat, St Sampsons, Guernsey, CI, UK

R A Wilson M/C
15 Willington Road, Kirton, Baston, Lincolnshire, UK

Tel: 0205 722282

R F Linton & Son
31 Springwell Street, Ballymena, Co.Antrim, UK

Tel: 0266 652516

Rawhide
Unit 9, Buxton Court, Manners Ind. Estate, Ilkeston, Derbyshire, DE7 8EF, UK

Tel: 0602 301555

Range of quality leathers - one year guarantee on all jackets and jeans. For fast delivery ring 0602-301555, or -443072, or -443074 Monday to Friday, 10am to 5pm.

Renham Motorcycles
The Parade, Thorpe Road, Staines, Middlesex, TW18 3HP, UK

Tel: 0784 458578

Renntec
69 Woolsbridge Ind.Estate, Three Legged Cross, Wimbourne, Dorset, BH21 6SP, UK

Tel: 0202-826722

Rolling Thunder UK
406 Staines Road, Bedfont, Middlesex, TW14 8BT, UK

Tel: 0629 57901

Shirlaws Motorcycles
92 Crown Street, Aberdeen, Gram- pian, AB1 2HJ, UK

Contact: Roy Shirlaw
Tel: 0224 584855

Mail order on bikes by BMW, Ducati, Triumph, Yamaha, Suzuki, Honda, Vespa. Parts, clothing & helmets, specialist repairs, training. Motorcycle world all under one roof.

Shoshoni Clothing
Atlantic Business Centre, Broadheath, Altrincham, Cheshire, WA14 5NQ, UK

Tel: 061 929 1149
Fax: 061 926 9520

Manufacturer and retailer of Shoshoni protective clothing - including the well proven Shoshoni Jeans. Phone or write for your free information pack. Riders welcome to call in.

Sondel Sports
28/32 Highbury Corner, London, N5 1RD, UK

Tel: 071 700 0310

Speedway Motors
78A Oldbury Road, Blackheath, Rowley Regis, Warley, Midlands (W.), B65 0JS, UK

Tel: 021 559 1270

Spellbound
46 Wood Street, Taunton, Somerset, TA1 1UW, UK

Tel: 0823 324516

Startline Accessories
504-507 Woodborough Road, Mappey, Nottinghamshire, UK

Startline Motorcycle Accessories
70-72 Sneinton Hermitage, Sneinton, Nottinghamshire, NG2 4BS, UK

Tel: 0602 412226

Stevens & Stevens Trials Centre
Unit 43,Blue Chalet Ind. Park, London Road, West Kingsdown, Kent, TN15 6BQ, UK

Tel: 0474 854265
Fax: 0474 854032

Trials motorcycles, spares, accessories & clothing. Advice on joining clubs (adults & juniors).

Taybike Clothing
31 Milnbank Road, Dundee, Tayside, UK

Tel: 0382 68000

Taylor Racing
23-25 Station Hill, Chippenham, Wiltshire, UK

Tel: 0249 657575/6

Tyre Sales Motorist Centre
1a Kingsholm Road, Gloucester, Gloucestershire, UK

Tel: 0452 525991

VMCC Regalia
112 Fairfield Crescent, Newhall, Swadlincote, Derbyshire, DE11 0TB, UK

Tel: 0283 224841

Victor Devine & Co Ltd
234 Gt.Western Road, Glasgow, Strathclyde, UK

Tel: 041 332 6264

Warmfit Ltd
1 Forester Road, Portishead, Bristol, Avon, BS20 9UP, UK

Tel: 0275 897570

Webbs of Lincoln
117-121 Portland Street, Lincoln, Lincolnshire, LN5 7LG, UK

Tel: 0522 528951

York Yamaha Centre
Haworth Village Garage, York, Yorkshire (N.), UK

Tel: 0904 424597

CLUBS & SOCIETIES

59 Club Classic Section
8 Badric Court, Yelverton
Road, Battersea, London,
SW11 3SW, UK

Contact: Mrs A Mooney

80cc Road Racing Club of GB
74 Ullswater Road,
Congleton, Cheshire,
CN12 4JJ, UK

Contact: Mr D Boothroyd

AJS & Matchless Owners Club
12 Chilworth Gardens,
Sutton, Surrey, SM1 3SP,
UK

Contact: Mr T Corley

AJS & Matchless Owners club
has 3000 + members worldwide,
monthly magazine, spares
scheme, rallies, 36 sections.

Aberdeen & District Moto Cross Club
20 Westdyke Drive, Elrick,
Skene, Aberdeen,
Grampian, AB32 6DR, UK

Advanced Motorcyclists Assoc. UK
44 Brook Road, Cadbury,
Warley, Midlands (W.),
B68 8AA, UK

Contact: A P Walsh

Amateur Motorcycle Association
Darlaston Road, The Pleck,
Walsall, Midlands (W.),
WS2 9XL, UK

Contact: Carol Davis

American Motorcycle Register
74 Reading Road,
Pangbourne, Berkshire, UK

Contact: Sheila Ward

Arena Essex Speedway
Arena Essex Raceway,
A1306 Arterial Road,
Purfleet, Essex,
RM16 1NX, UK

Contact: Peter Thorogood
Tel: 0268 753067

Ariel Owners Club
5 The Meadows, Yeland
redmayne, Carnforth,
Lancashire, LA5 9SY, UK

Contact: Denys Wilkinson

Army Motorcycling Association
A.S.M.T., Normandy
Barracks, Leconfield,
Humberside (N.),
HU17 7LX, UK

Contact: Lt Col E P Bartlett

Assoc. of Pioneer Motor Cyclists
"Sulby", Sparkford Hill
Lane, Sparkford, Nr Yeovil,
Somerset, BA22 7JF, UK

Contact: Jeff Clew
Tel: 0935 850605

Holding meetings, rallies and
dinners where the Association's
memorabilia is displayed.
Minimum qualification for
membership: To have held a
motorcycle driving licence for 40
years.

Association Of Independent Motorcyclists
322 Grangemouth Road,
Radford, Coventry,
Midlands (W.), CV6 3PL,
UK

Contact: Brian Clarke

Association Of Motorcyclists Against Discriminatory Legislation
18 Johnston Road, Oakdale,
Poole, Dorset, BH15 3NT,
UK

Contact: Ivan Squires

Association of Hill Climb Clubs
505a Penistone Road,
Shelly, Huddersfield,
Yorkshire (W.), UK

Contact: Mr H Halstead

Association of Road Race Clubs
12 Mordaunt Drive, Four
Oaks, Sutton Coldfield,
Midlands (W.), B75 5PT,
UK

Contact: Mike Dawes

Association of Sporting Referees
64 Lowton Road, Colborne,
Warrington, Cheshire, UK

Contact: Mr J Miller

Auto-Cycle Union
Miller House, Corporation
Street, Rugby,
Warwickshire, CV21 2DN,
UK

Automotive Manufacturers Racing Assoc.
c/o Premier Fuel Systems
Ltd, Wilow Road, Castle
Donnington, Derbyshire,
DE7 2NP, UK

Tel: 0332 850515

Avon Valley MCC
Woodside House, Ashgill,
Larkhall, Strathclyde, UK

Tel: 0698 881962

BMF
35 Gellantly Place, Brechin,
Angus, Tayside, DD9 6BS,
Scotland

Contact: Stewart Mowatt
Tel: 0356 623981

BMF
13 South Terrace, Sowerby,
Thirsk, Yorkshire (N.),
YO7 1RH, UK

Contact: John Carrington
Tel: 0845 524264

BMF
33 Park Lane, Ramsden
Heath, Billericay, Essex,
CM11 1NE, UK

Contact: Graham Butler
Tel: 0268 710125

BMF
Ladybower, Dogmersfield,
Basingstoke, Hampshire,
RG27 8SS, UK

Contact: Don Lewis
Tel: 0252 616359

BMF
Tremore, Liskey Hill
Crescent, Perranporth,
Cornwall, TR6 OHP, UK

Contact: Derek Crofts
Tel: 0872 573537

BMF
18 Roundwood Close,
Cyncoed, Cardiff,
Glamorgan (S.), CF3 7HH,
UK

Contact: Dave Jones
Tel: 0222 494719

BMF
12 Nelson Street,
Audenshaw, Manchester,
Gt., M34 5EF, UK

Contact: Pete Crowe
Tel: 061 320 4597

BMF
Brewhouse, Redhall,
Ballycarry, Carrickfergus,
Co.Antrim, BT38 9JL, UK

Contact: Irene McClintock
Tel: 096 03 72373

BMF
20 Church Street,
Ringstead,
Northamptonshire,
NN14 4DH, UK

Contact: *Graham Sanderson*
Tel: 0933 625176
Fax: 0933 625584

BMF
49 Sarratt Avenue, Hemel
Hempstead, Hertfordshire,
HP2 7JN, UK

Contact: *Trevor Magner*
Tel: 0442 246944

BMF
101 Square Lane,
Burscough, nr Ormskirk,
Lancashire, L40 7RG, UK

Contact: *Tim Stevens*
Tel: 0704 894136

BMF
67 New End, London,
NW3 1HY, UK

Contact: *Stephen Prower*
Tel: 071 431 3819

BMF
90B Monega Road, Forest
Gate, London, E7 8EW,
UK

Contact: *Sharon Nash*
Tel: 081 470 4441

BMF
Midlands Region, 19 Tate
Grove, Hardingstone,
Northamptonshire,
NN4 6UY, UK

Contact: *Stephen Bergman*
Tel: 0604 709771

Protecting and promoting motor
cycling. For membership details,
individual or club, events and
meetings, write including SAE.
Motorcyclists working for
motorcyclists.

BMF Rally
75 Western Road,
Leicester, Leicestershire,
LE3 0GE, UK

Contact: *Mike Fairhead*
Tel: 0533 548818
Fax: 0533 543030

BMF Rally, East of England
showground, Peterborough.
Europe's largest outdoor
motorcycle show, May 22nd
1994. Opens 9am: Public £6.00,
members £5.00, under 14's - Free.

BMF Rider Training
Scheme
PO Box 2, Uckfield, Sussex
(E.), TN22 3ND, UK

Tel: 0825 712896
Fax: 0825 712787

BMW Club
22 Combermere,
Thornbury, Bristol, Avon,
BS12 2ET, UK

Contact: *Mike Cox*
Tel: 0454 415358

National club of BMW motor
cycle owners with nearly 6000
members. Large monthly
magazine. Extensive country
wide social calendar. Help
available for new and old bikes.

BSA Bantam Racing
Club
6 Kipton Close, Rothwell,
Northamptonshire, UK

Contact: *Mrs J Walpole*

BSA Bantam Enthusiasts
Club
5 Surrey Way, Gillingham,
Kent, ME8 6XB, UK

Contact: *Dave Burton*
Tel: 0634 379235

Catering for Bantams, Bugles,
Dandys and Winged Wheels,
wide range of services including
reproduction literature,
bi-monthly magazines and
discount spares scheme. Send
SAE for details.

BSA Owners Club
44 Froxfield Road, West
Leigh, Havant, Hampshire,
PO9 5PW, UK

Contact: *Rob Jones*

Backpackers Club
7-10 Friar Street, Reading,
Berkshire, RG3 4RL, UK

Backpackers Club
20 St.Michael Road,
Tilehurst, Reading,
Berkshire, RG3 4RP, UK

Bantam Grasstrack
Association
10 Duchess Drive,
Bridgnorth, Shropshire,
WV16 4JD, UK

Contact: *Paul Bearman*

Barn Hill Farm
Whaddon, Milton Keynes,
Buckinghamshire,
MK17 0NQ, UK

Contact: *D B Lowe*
Tel: 0908 501750

Belle Vue Speedway
Greyhound Stadium,
Kirkmanshulme lane,
Gorton, Manchester, Gt.,
M18 7BA, UK

Contact: *John Perrin*
Tel: 061 736 0128

Benelli (Brembo Club
International)
21 Mount Pleasant, Sutton
in Ashfield,
Nottinghamshire, UK

Contact: *P Rimmer*

Berkeley Enthusiasts
Club
41 Gorsewood Road, St
Johns, Woking, Surrey,
GU21 1UZ, UK

Contact: *M Rounsville Smith*

Birmingham MCC
26 Fieldon Close, Shirley,
Solihull, Midlands (W.),
B90 3ED, UK

Contact: *Mr G Harrison*
Tel: 021 745 2557

Bon Accord MCC
3 Woodview Place,
Stonehaven, Grampian,
AB3 2GD, UK

Contact: *Diane Stuart*
Tel: 0569 63025

Bradford Speedway
Odsal Stadium, Rooley
Avenue, Bradford,
Yorkshire (W.), BO6 1BS,
UK

Contact: *Allan Ham*
Tel: 0274 690614

Brighton & District MCC
73 Eastbrook Road,
Portslade, Sussex (E.),
BN41 1P3, UK

Contact: *Ian Swyer*

Brighton & District MCC is the
longest established club on the
South Coast. Meets at the
Southern Cross Club, Victoria
Road, Portslade, East Sussex. All
social and sports activities
catered for. BMF and ACU
affiliated. Contact Ian Swyer on
0273 430458.

British Drag Racing
Association
29 West Drive, Highfield,
Caldecote, Cambridgeshire,
CB3 7NY, UK

Contact: *Mrs Y Tramm*

British Formula Racing
Club
Cronk Y Voddy, Rectory
Road, Coltishall, Norwich,
Norfolk, UK

Contact: *J Milligan*

British M/C Racing
Marshals Assoc.
Brands Hatch Circuit,
Fawkham, Dartford, Kent,
DA3 8NG, UK

Contact: *Terry Mount*

British MC RC
Brands Hatch Circuit,
Fawkham, Dartford, Kent,
DA3 8NG, UK

Contact: *Terry Mount*

British Motor Cycle Club
Kimton, Les Grippios,
Bordeaux, Vale, Guernsey,
CI, UK

Contact: *Tony Scowen*

British Motorcycle
Owners Club
59 Mackenzie Street,
Bolton, Lancashire, UK

Contact: *Phil Coventry*

British Motorcycle Preservation Society
Flat 3, 41 Brailsford Court, Brailsford Road, Ecclesfield, Sheffield, Yorkshire (S.), S5 9DJ, UK

Contact: Paul Thompson

British Motorcycle Riders Club (Oxford)
PO Box 2, Eynsham, Witney, Oxfordshire, OX8 1RW, UK

Contact: Geoff Ives
Tel: 0865 880626

A club for riders of British motorcycles. We meet at the Chequers Public House in Cassington (off A40 between Oxford and Witney) every Monday night.

British Schoolboy Motorcycle Assoc.
18 Glenpark Crescent, Kingscourt, Stroud, Gloucestershire, GL5 5DT, UK

Contact: Mrs L Hill

British Speedway Promoters Assoc
A.C.U.House, Wood Street, Rugby, Warwickshire, CV21 27X, UK

Tel: 0788 560648
Fax: 0788 546785

British Two Stroke Club
38 Charles Drive, Cuxton, Kent, ME2 1DR, UK

Contact: Mr A Abrahams

Brothers Of The Third Wheel
5 Kingsbridge Court, Harbour Way, Folkestone, Kent, UK

Brough Superior Club
Flint Cottage, St Pauls Walden, Hitchin, Hertford-shire, SG4 8DN, UK

Contact: Justin Wand

The club organises runs and rallies throughout the year, with an annual rally in August. Monthly newsletter and active spares scheme. Membership details from secretary.

Bryn Haulwen
Maesllyn, Llandysul, Dyfed, SA44 5JP, UK

Contact: Neil Davis
Tel: 0239 851729

Bucks British Motorcycle Club
The Hampden Arms, Great Hampden, Great Missenden, Buckinghamshire, HP16 9RQ, UK

Contact: Liz Wiltshire
Tel: 0494 714864

"BBMCC" is a social club for owners, restorers and riders of British motorcycles, meeting every Wednesday night at The Hampden Arms, Great Hampden, Great Missenden.

CBX Riders Club (UK)
57 Osborne Road, Basingstoke, Hampshire, RG21 2TS, UK

Contact: Peter Broad

CSMA
29 First Avenue, Church Accrington, Lancashire, BB5 5EH, UK

Contact: Mr S Johnson
Tel: 0254 392663

CSMA (East Midlands)
8 Forsythia Close, Branston, Lincolnshire, LN4 1PS, UK

Contact: Mr T Kirby
Tel: 0522 791546

CSMA (Midland)
67 White Street, Balsall Heath, Birmingham, Midlands (W.), B12 8SF, UK
Contact: Mr W D Francis

CSMA (East Yorks)
20 Hodgson Lane, Drighlington, Bradford, Yorkshire (W.), BD11 1BN, UK

Contact: Mr D Smith
Tel: 0532 852805

CSMA
Brittania House, 95 Queens Road, Brighton, Sussex (E.), BN1 3WY, UK

Contact: Mr Devenish
Tel: 0273 21921

Callander MCC
Ardmar, 70 Main Street, Callander, Central, FK17 8BD, UK

Contact: Rod Johnson
Tel: 0877 31068

A social club organising runs, rallies, dances etc. S.A.C.H., BMF affiliated, membership £15 p.a. Road riders and racers welcome. Meetings held alternative Wednesdays.

Camel Vale MCC
1 Halgavor Road, Bodmin, Cornwall, PL31 1BW, UK

Contact: S.Madgwick
Tel: 0208 74411

ACU affiliated club running a range of trials, enduros, classic trials and social events.

Cheshire MC Marshals Association
49 Woodsmoor Lane, Woodsmoor, Stockport, Cheshire, SK2 7AZ, UK

Contact: Mr R N Richards
Tel: 061 456 4941

Civil Service MCC
Britannia House, 95 Queens Road, Brighton, Sussex (E.), BN1 3WY, UK

Contact: Mr C Devenish

Civil Service Motoring Association
Pollards, Badingham, Woodbridge, Suffolk, IP13 8LU, UK

Contact: Mr P Collins
Tel: 072 875217

Classic Racing MCC
Fenn Farm, St Marys Ho, Rochester, Kent, MN3 8QY, UK

Contact: Anne Murden

Cleeton Court Farm
Cleeton St Mary, Cleobury Mortimer, Kidderminster, Hereford & Worcester, DY14 OQZ, UK

Contact: Cliff Pearce
Tel: 058 475 354

Club Yamaha
PO Box 27f, Chessington, Surrey, KT9 1SH, UK

Congleton & DMC
18 Springfield Drive, Congleton, Cheshire, UK

Contact: Mr M Beech
Tel: 0260 275381

Cossack Owners Club
19 Elms Road, Bare, Morecambe, Lancashire, LA4 6AP, UK

Contact: Mr Phil Hardcastle

Coventry Speedway
Coventry Stadium, Rugby Road, Brandon, Nr Coventry, Warwickshire, CV8 3GJ, UK

Contact: Martin Ochiltree
Tel: 0203 542395

Cradley Heath Speedway
Cradley Heath Stadium, Dudley Wood Road, Dudley, Midlands (W.), DY2 ODH, UK

Contact: Colin Pratt
Tel: 0384 635700

Crusader Bike Tours Ltd
68-69 The Mint, Rye, Sussex (E.), UK

Tel: 0797 224640

Cwm Derw
Pant y dwr, Rhayader, Powys, LD6 5NB, UK

Contact: Tommy Breeze
Tel: 059788 202

DKW Rotary Owners Club
Dunbar, Ingatestone Road, Highwood, Chelmsford, Essex, CM1 3QU, UK

Contact: *David Cameron*

Already with a section for Suzuki RE-5 owners, the club is starting to attract rotary Nortons thus becoming more of a general "Wankel Motorcycle" organisation.

DOT Owners Club
4 Pendragon Place, Failsworth, Mancheser, Gt., M35 9GL, UK

Contact: *Eric Watson*

Dalmellington MCC
12 Main Street, Sorn, Ayrshire, UK

Contact: *James Morton*
Tel: 0290 51353

Donington 100 MCC Ltd
21 Tamworth Street, Duffield, Derbyshire, DE3 4ER, UK

Contact: *Mrs Smith*
Tel: 0332 842019

Donnington Supporters Club
80 Baldocks Lane, Melton Mowbray, Leicestershire, LE13 1EW, UK

Contact: *Tony Gamble*
Tel: 0664 64042

Club benefits include gate concessions, free paddock transfer, private enclosure with covered grandstand and portakabin, newsletters, circuit rides. Membership costs just £11 single, £17 joint.

Ducati 900SS Riders Co.Ltd
10 Cambrai Road, Chiseldon, Swindon, Wiltshire, SN4 0JD, UK

Ducati Owners Club
131 Desmond Drive, Old Catton, Norwich, Norfolk, NR6 7JR, UK

Contact: *Sue Purdy*

Dunfermline & District MCC
19 Dean Ridge, Gowkhall, Dunfermline, Fife, KY12 9PE, UK

Contact: *Alan Gibson*
Tel: 0383 851288

Dunmow & DMCC Ltd
6 Beaumont Hill, Great Dunmow, Essex, CM6 2AP, UK

Contact: *Mr Julian Sayer*
Tel: 0371 874756

Affiliated to Eastern centre of The Auto-Cycle Union. Club caters for adult off-road events, particularly grass track, moto-cross and trials. Events organised for 1994 include: Scramble - Stebbing, Essex 01/05/94. Grass track - Ugley, Stansted, Essex 19/06/94 & 18/09/94. Membership enquiries to Dean Lambert, Tel:0245-441072.

Dunstall Owners Club
16 Agecroft Road West, Prestwich, Manchester, Gt., UK
Contact: *B Hutchinson*

East Lothian Road & Trail Club
4 Beanston Mains, Haddington, Lothian, UK

Contact: *Michael Foy*
Tel: 0620 88284

East Midland Scooter Association
9 Coronation Street, Mansfield, Nottinghamshire, NG18 2QL, UK
Contact: *Carol Braithwaite*

Eastbourne Speedway
Arlington Stadium, Arlington Road West, Hailsham, Sussex (E.), BH27 3RE, UK
Contact: *John Crook*
Tel: 0273 454555

Eastern Centre Marshals Club
83 Alexandra Drive, Wivenhoe, Colchester, Essex, CO7 9SF, UK
Contact: *Mr D Lawrence*
Tel: 0206 824512

Edinburgh & District MC Ltd
28 Nelson Street, Edinburgh, Lothian, EH3 6LJ, UK

Contact: *J. McColm*
Tel: 031 556 1031

Edinburgh Southern MC
18/1 Ladynairn Loan, Edinburgh, Lothian, EH8 7NN, UK

Contact: *R Adamson*
Tel: 031 661 1672

Edinburgh Speedway
Powerhall Stadium, Beaverhall Road, Edinburgh, Lothian, EH7 4JE, UK

Contact: *John Campbell*
Tel: 031 663 1575

Edinburgh St George MC
Blinkbonny Farm Cottages, Gorebridge, Midlothian, UK

Contact: *Anne Gordon*
Tel: 0875 21326

Egli Vincent Owners Club
17a Picket Piece, Andover, Hampshire, SP11 6LY, UK

Contact: *P French*

Elsworth Moto Park
34 High Street, Melbourn, Hertfordshire, SG8 6DZ, UK

Contact: *Andy Lee*
Tel: 0763 260658

End to End Endurance Club of GB to End Endurance Club of GB
2 Kendal Bank, Leeds, Yorkshire (W.), LS3 1NR, UK

Contact: *Ken Kirk*

Essex Trials Group
25 Seabrook Gardens, Boreham, Chelmsford, Essex, CM3 3BX, UK

Contact: *Mr T J Woodmason*
Tel: 0245 466815

Exeter Speedway
County Ground Stadium, Cowick Street, St Thomas, Exeter, Devonshire, UK
Contact: *Colin Hill*
Tel: 0392 50956

Fantic Owners Trail Club
White Lodge, Blue Bell Hill, Aylesford, Kent, UK
Contact: *Mr R Cooper*

Featherbed Specials Owners Club
Maytham Farm, Hatters Hill, Rolvenden, Cranbrook, Kent, TN17 4QA, UK

Launched in April 1992 and strictly dedicated to the construction and interest of all "Featherbed" based specials. Membership - 150. Activities include a newsletter, runs and club-stands.

Federation of British Scooter Clubs
219 Elmers End Road, Beckenham, Kent, BR3 4EH, UK
Contact: *Mrs Sylvia Hillman*

Federation of European Motorcyclists
45 St Katherines Road, Whipton, Exeter, Devonshire, EX4 7JW, UK

Federation of Sidecar Clubs
1A Mowbrays Road, Romford, Essex, RM5 3ET, UK
Contact: *Phil Burnham*

Fedn. of British Police Motorcycle Clubs
PS 5250 Cairns, Traffic Office, Petersfield, Hampshire, GU32 3HU, UK

Fellowship of Christian Motorcyclists
Flat 4, Kingsbridge, 172 Lordship Road, London, N16 5HB, UK
Contact: *Richard Edelsten*

If you find secular clubs too worldly, try us! Regular outings, no formal outreach, the key word is fellowship. National club.

Fellowship of Historic Motorcycle Assocs
75 Boulton Grange, Randlay, Telford, Shropshire, TF3 2LD, UK

Contact: John Law

Forgotten Racing Club
73 High Street, Morton, Bourne, Lincolnshire, PE10 ONR, UK

Contact: Chris Pinches

Francis Barnet Owners Club
Whitehorse Cottage, White Horse Lane, Sutton Poyntz, Weymouth, Dorset, DT3 6LU, UK

Contact: Keith Young

GRASA
27 Aylesbury Way, Yateley, Camberley, Surrey, GU17 7NS, UK

Contact: Mr R Cross

GT/GTR Kawasaki Owners Club
100 Belville Street, Fenton, Stoke on Trent, Staffordshire, ST4 3LH, UK

Galloway MCC
Cowthat, Ecclefechan, Lockerbie, Dumfries & Galloway, UK

Contact: Janice Forish
Tel: 0576 300242

Garnock Academy MCC
Physics Department, Garnock Academy, Kilbirnie, Strathclyde, UK

Contact: P Urquhart
Tel: 0505 682658

Gay Bikers MCC
c/o Gay Switchboard, 31a Mansfield Road, Nottingham, Nottinghamshire, NG1 3FF, UK

Contact: The Secretary

Gilera Information Exchange
Fox House, Moor Road, Langham, Colchester, Essex, CO4 5NR, UK

Contact: Gerard Gilligan
Tel: 0206-272737

The Gilera Information Exchange exists as a central point for obtaining information on all aspects of the Gilera Marque. Free.

Glasgow Speedway
Shawfield Sports Stadium, Rutherglen Road, Glasgow, Strathclyde, G73 1SZ, UK

Contact: Neil Macfarlane
Tel: 041 613 1552

Glenrothes Youth MCC Ltd
The Cottage, Victoria Road, Ladybank, Fife, KY7 7LJ, UK

Contact: Pauline Lawson
Tel: 0337 31177

Gold Wing Owners Club of GB
18 Arncliffe Way, Cottingham, Humberside (N.), HU16 5DH, UK

Contact: J D Horner

Goldring Barn Raceway
Henfield Road, Small Dole, Sussex (W.), UK

Tel: 0903 816236

Grampian MCC
17 Blackfriars Road, Elgin, Grampian, IV30 1TY, UK

Contact: Mark Watson
Tel: 0343 552794

Grass Track Riders Club
13 Greetham Road, Aylesbury, Buckinghamshire, UK

Contact: Mr M Surman

Grass Track Sidecar Association
The Onslow Arms, High Street, Loxwoods, Sussex, UK

Contact: Miss L O'Hare

Greeves Riders Association
4 Longshaw Close, North Wingfield, Chesterfield, Derbyshire, S42 5QR, UK

Contact: David McGregor

Formed in 1983 by Greeves enthusiasts, to provide a club for Greeves owners, to contact each other and help organise a parts and information exchange.

Hackpen Farm
Wroughton, Swindon, Wiltshire, UK

Contact: Shane
Tel: 0793 812224

Halstead & DMCC
6 Alan Way, Lexden, Colchester, Essex, CO3 4LG, UK

Contact: Mr A C Adams
Tel: 0206 540484

Adult only - four meetings a year at Little Loveney Hall, Wakes Colne, Essex. Two British Moto-Cross - two Eastern centre championship events. Membership £16.00.

Harley-Davidson Riders Club
PO Box 62, Newton Abbott, Devonshire, TQ12 2QE, UK

Contact: Mitzi Belsher

H.D. & Indian enthusiasts:- Solos, 3 wheels, vintage, singles, insurance, products, rallies, discounts, FEM, BMF, MAG, tool hire, archives. Bi-monthly colour magazine. Est.1949.

Harris Owners Club
18 Corsham Road, Fords Farm, Calcott, Reading, Berkshire, RG3 5ZH, UK

Contact: Keith Fothergill

Harrison Scrambling Track
Willenhall Lane, Bloxwich, Walsall, Midlands (W.), UK

Tel: 0922 710670

Hawick & Borders MCC
Sunnyside, Denholm, Hawick, Borders, UK

Contact: Mr J Steel
Tel: 0450 87505

Hedingham Sidecar Owners Club
14 Court Road, Burham, Rochester, Kent, ME1 3TA, UK

Contact: Alan Critcher
Tel: 0634 863132

Membership open to past, present & future owners of Hedingham sidecars. Club magazine, newsletter, rallies, club runs & social events.

Heinkel/Trojan Owners/ Enthusiasts Club
Carisbrooke, Wood End Lane, Fillingley, Coventry, Midlands (W.), CV7 8DF, UK

Contact: Y Luty

Hesketh Owners Club
1 Northfield Road, Soham, Cambridgeshire, CB7 5UE, UK

Contact: P White

Honda CBX 550 Owners Club
14 The Ridings, Bishopsworth, Bristol, Avon, BS13 8NV, UK
Contact: Tim Pursall

Honda Gold Wing Owners Club
796 Central Hill, Crystal Palace, London, SE19 1BS, UK

Contact: May Gambrill

Honda Owners Club (GB)
18 Embley Close, Calmore, Southampton, Hampshire, SO4 3QX, UK

Branches nationwide; regular magazine, dealer discounts, classic selection, club runs, rallies, camping weekends, technical, regalia, active social life; founded in 1961.

Humberside County Council MCC
Hallgarth, 120 King Street, Cottingham, Humberside (N.), UK

Contact: Clive Piper
Tel: 0482 847501

Ilford Amateur MC
16 Pulleyns Avenue, East Ham, London, E6 3LZ, UK

Contact: Mr G Parkins
Tel: 081 472 8441

Ilford MC & LCC
110 Cranham Road, Hornchurch, Essex, RM11 2BA, UK

Contact: Mr K Price
Tel: 04024 55525

Indian Owners Motorcycle Club of GB
183 Buxton Road, Newtown, New Mills, Stockport, Cheshire, SK12 3LA, UK

Contact: J.Chatterton
Tel: 0663 747106

The club caters for owners of Indians and other rare American bikes (Henderson, Pope etc). Parts scheme, annual owners rally plus regular magazines. Membership fee £15.00.

Institute of Motorcycling
Cowley House, 9 Little College Street, London, SW1E 3XS, UK

Tel: 071 222 0664
Fax: 071 233 0335

International Christian Classic Motorcyclists Association
The Rectory, Nuffield, Henley on Thames, Oxfordshire, RG9 5SB, UK

Contact: John Shearer

International Four Stroke Owners Club
197 Northwood Road, Thornton Heath, Surrey, CR4 8HX, UK

Contact: Miss K Ableman

International Laverda Owners Club
21 Wansbury Way, Swanley, Kent, BR8 8DH, UK

Contact: Nick Clements
Tel: 0322-664314

The Laverda Owners Club (established 1974, with 1,000 members) offers club magazine, track-days, insurance discounts, regalia, technical advice, local meetings, rallies and much, much more.

International Motor-Cycle Group
18 Lambo Park, Maidstone Road, Paddock Wood, nr Tonbridge, Kent, TN12 6DA, UK

Contact: Colin Packman

International Motorcyclists Tour Club
Hazels, Orchard Close, Burrow Lane, Newton Poppleford, Devonshire, EX10 0BB, UK

Contact: Mrs Janet Jenkins

Inverness & District MCC
23 Ardholm Place, Inverness, Highland, UK

Contact: Mr J MacLennan
Tel: 0463 712149

Ipswich Speedway
Foxhall Stadium, Foxhall Road, Ipswich, Suffolk, IP4 5TL, UK

Contact: John Lois
Tel: 0473 749288

Isle of Wight British Motorcycle Soc.
73b Blythe Way, Shanklin, Isle of Wight, PO37 7NL, UK

Contact: Alan Wilson

Italian Motorcycle Owners Club (GB)
14 Rufford Close, Barton Seagrave, Kettering, Northamptonshire, NN15 6RF, UK

Contact: Rosie Marton

Jawa/CZ Owners Club
Old Dairy, Achmaha by Kilchoan, Argyll, Highland, PH36 4LW, UK

Contact: Tony Thain

Joyce Green Motocross Site
54 Holbrook Way, Bromley, Kent, BR2 8EE, UK

Kawasaki Triples Club
132 Bells Lane, Cinderhill, Nottingham, Nottinghamshire, NG8 6DW, UK

Contact: Rick Brett
Tel: 0602-274777

The Kawasaki Triples Club has over 300 U.K. members and several sister clubs worldwide, it is invaluable for locating new and used parts and information when restoring a Triple.

Kawasaki Z1 Owners Club
54 Hawthorne Close, Congleton, Cheshire, CW12 4UF, UK

Contact: Sam Holt

Kelvedon Hatch Trials Club
Gt Myles, Kelvedon Hatch, Brentwood, Essex, CM15 0LB, UK

Tel: 0277 372562

Kings Lynn Speedway
The Stadium, Saddlebow Road, Kings Lynn, Norfolk, PE34 3AG, UK

Contact: Keith Chapman
Tel: 0553 771111

Kinlochleven Motorcycle Club
Mamore, Kinlochleven, Highland, UK

Contact: George Louden
Tel: 08554 337

Kinross & District MCC Ltd
Strone, The Square, Letham, Fife, KY7 7RP, UK

Contact: Atholl Cameron
Tel: 0337 81370

Kirkcaldy & District MC Ltd
9 Duddingstone Drive, Kirkcaldy, Fife, KY2 6JP, UK

Contact: Ian Barnes
Tel: 0592 266902

Knockhill MCRC Ltd
Knockhill Racing Circuit, Dunfermline, Fife, KY12 9TF, UK

Contact: Binkie Chapman
Tel: 0383 731788/723337

Lambretta Club GB
8 Trent Close, Rainhill, Prescot, Merseyside, L35 9LD, UK

Contact: Kevin Walsh
Tel: 051 426 9839

Everything for the Lambretta enthusiast.

Lambretta Preservation Society
Kesterfield, North Lew, Okehampton, Devonshire, UK

Contact: Mike Karslake

Langbaurgh Motorsports Park
Middlesbrough Road, South Bank, Teeside, Yorkshire (N.), TS1 6EL, UK

Contact: Jeff Sadler
Tel: 0642 252555

Le Velo Club
146 Hall Lane, Walsall Wood, Walsall, Midlands (W.), WS9 9AP, UK

Contact: Colin Roberts

Le Velo Club Limited
Chapel Mead, Winterborne, Whitechurch, Blandford, Dorset, DT11 0AB, UK

Contact: Kevin Parsons

Le Velo club caters for Le Valiant, Vogue and Viceroy made from 1948 to 1971. Magazine, spares and regular meetings.

Lickey Ash MCC Ltd
Tulyar, 34 Elm Grove,
Norton, Bromsgrove,
Hereford & Worcester,
B61 0EH, UK

Contact: Mr L Plews
Tel: 0527 575634

Loch Fyne Motorcycle Club
47 MacIntyre Terrace,
Lochgilphead, Strathclyde,
PA31, UK

Contact: M Harvey

Lochaber & District MCC
20 Mulroy Terrace, Roy
Bridge, Highland, PH31
4AF, UK

Contact: Grace Dignan
Tel: 0397 81 268

London Area Scooter Clubs Association
409 Wickham Road,
Shirley, Croydon, Surrey,
CR0 0DP, UK

Contact: Mr M Tctum

London Douglas MCC
48 Standish Avenue, Stoke
Lodge, Patchway, Bristol,
Avon, UK

Contact: Reg Holmes

Long Eaton Speedway
Station Road, Long Eaton,
Nottinghamshire,
NG10 2DU, UK

Contact: Eric Boocock
Tel: 0924 463435

Lothian & Borders C&VMCC
Winton Cottage, Polton
Village, Lasswade
Midlothian, UK

Contact: Hamish Buchanan

MV Agusta Owners Club
31 Baker Street, Stapenhill,
Burton on Trent,
Staffordshire, DE15 9LU,
UK

Contact: Martyn J Simpkin
Tel: 0283 46535

To provide information, spares,
regalia and a regular MV
magazine to all club members
around the world.

MV Agusta Riders Club GB
27 Melbourne Road,
High Wycombe,
Buckinghamshire,
HP13 7HE, UK

Contact: Richard Boshier

MZ Riders Club
14 Victoria Place, Stoke on
Trent, Staffordshire,
ST4 2LU, UK

Contact: A W Gotham

Maico Owners Club
"No Elms", Goosey,
Farringdon, Oxfordshire,
SN7 8PA, UK

Contact: Phil Hingston
Tel: 0367-710408

The Maico Owners Club is
dedicated to the preservation of
machines circa 1950 - 1967,
producing spares, and
newsletters to keep enthusiasts
in touch world wide.

Manchester Eagles MCC
263 Wythenshawe Road,
Wythenshawe, Manchester,
Gt., M23 9DE, UK

Tel: 061 998 5093

Manchester Eagles MCC
50 Harley Road, Sale,
Cheshire, M33 1FP, UK

Contact: Mr M J Biggs

Manmoel Motocross Practice Track
1 Philip Street, Nr Crumlin,
Newport, Gwent, NP1 4JF,
UK

Contact: C Watkins

Manx Grand Prix Riders Association
Windy Corner, 3 Chosen
Drive, Churchdown,
Gloucestershire, GL3 2QS,
UK

Contact: George Ridgeon

Market Harborough British M/C Club
50 Weir Road, Kibworth
Beauchamp, Leicestershire,
UK

Contact: Mrs Edna Bowden

Marston Sunbeam Register
Marston Palmer Limited,
Wobaston Road,
Fordhouses,
Wolverhampton, Midlands
(W.), UK

Contact: Cyril J Wakeman

Matchams Motorcycle Club
37 Church Road, Hayling
Island, Sussex (W.),
PO11 0NN, UK

Contact: Mr Slack
Tel: 0705 462282

Melville MC (Scot) Ltd
9 Carlops Road, Penicuik,
Midlothian, EH26 0EY, UK

Contact: Tom Dickie
Tel: 0968 673611

Messerschmitt Owners' Club
21 Fairford Close, Church
Hill North, Redditch,
Hereford & Worcester,
B98 9LU, UK

Contact: Alan Marriott

Messerschmitt Owners' Club
(founded 1956), the largest
Messerschmitt club in the world
with members in all continents.
Monthly magazine, spares,
rallies and social events.

Mid Cheshire MXC
9 Poplar Close, Coed
Poeth, Wrexham, Clwyd,
LL11 3LZ, UK

Tel: 0978 753620

Middle England Scooter Association
35 Pembroke Croft, Hall
Green, Birmingham,
Midlands (W.), B28 9EY,
UK

Contact: T L Sharp

Middlesbrough Speedway
Cleveland Park Stadium,
The Fossway, Byker,
Newcastle upon Tyne, Tyne
& Wear, NE6 2XJ, UK

Contact: Tim Swales
Tel: 0532 584611

Midland (Moto) Rumi Club
Woodbine Cottage, 232
Trysull Road, Merry Hill,
Wolverhampton, Midlands
(W.), WV3 7JR, UK

Contact: Ray Pearce

Midland Drag RA
27 Maple Road, Sutton
Coldfield, Midlands (W.),
B72 1JP, UK

Contact: Mr R Dobbie

Military Vehicle Conservation Group
Sundown, 83 New Road,
Broham, Chippenham,
Wiltshire, UK

Contact: Chris Orchard

Military Vehicle Trust
PO Box 6, Fleet,
Hampshire, GU13 9PE, UK

Morgan Three Wheeler Club Ltd
Arden Cottage, Rocky
Lane, Benllech, Anglesey,
Gwynedd, LL74 8TN, UK

Contact: Morris Blease

Morini Rides Club
1 Glebe Farm Cottages,
Sutton Veny, Warminster,
Wiltshire, BA12 7AS, UK

Contact: Kevin Bennett

The Morini Riders Club consists
of about 300 members nationally
and overseas. We offer our own
magazine and usual club
benefits, and are B.M.F. affiliated.

Moto Guzzi Club (GB)
53 Torbay Road, Harrow, Middlesex, UK

Contact: Penny Trengrove
Tel: 081 864 4922

For enthusiasts throughout the world. We offer: magazines, discounts, technical advice, social events, camping weekends, continental touring, factory visits, an insurance and valuation service and Europ Assistance.

Motocross Practice Site
Grafton Road, Brigstock, Corby, Northamptonshire, UK

Tel: 0536 373518

Motor Cycle Industry Assoc. of GB Ltd.
Stanley House, Eaton Road, Coventry, Midlands (W.), CV1 2FH, UK

Motorcycle Action Group
PO Box 750, King's Norton, Birmingham, Midlands (W.), B30 3BA, UK

Motorcycle Association of GB
Starley House, Eaton Road, Coventry, Midlands (W.), UK

Motorcycle Club of Cheshire
5 Cedar Close, Old Mill Estate, Bradley, Wrexham, Clwyd, LL11 4DL, UK

Contact: Mrs D Reaney
Tel: 0928 758535

Motorcycle Racing Sponsors Association
79 Sandfield Road, Arnold, Nottinghamshire, UK

Contact: Mr D Bingham

Motorcycle Retailers Association
201 Great Portland Street, London, W1N 6AB, UK

Tel: 071 580 9122

Motorcycling Club Ltd
Upper Stonecroft, Finmere, Buckingham, Buckinghamshire, MK18 4JA, UK

Contact: R Tucker Peake

Motozone
Bulls Head House, Nantwich Road, Audley, Stoke on Trent, Staffordshire, ST7 8DH, UK

Contact: Ron Carter
Tel: 0782 721922

NSU Register
5 Woodside Crescent, Clayton, Newcastle Under Lyme, Staffordshire, ST5 4BW, UK

Contact: Andrew Gibson

National Auto-Cycle and Cyclemotor Club
1 Parkfields, Roydon, Harlow, Essex, CM19 5JA, UK

Contact: Mr Rob Harknett
Tel: 0279 792329

The National Auto-Cycle & Cyclemotor Club, built over 25 years ago, is open to anyone interested in auto-cycles, cycle motors & mopeds. The club produces a bi-monthly magazine, has a library, spares department, transfer department & machine dating service.

National Bloodrunners Register
c/o SERV, The Malden Centre, Blagdon Road, New Malden, Surrey, UK

Contact: Dave Coppin

National Deaf Motorcycle Club
Royal Association for the Deaf, 27 Old Oak Road, Acton, London, W3 7HN, UK

National Drag Racing Association
27 Hillfields, Foxton, Cambridgeshire, UK

Contact: Mrs W Talbot

National Handicapped Motorcyclists Assoc.
20 Barlow Street, Radcliffe, Manchester, Gt., M26 9SU, UK

Contact: Tracy Connell

National Hillclimb Association
Piglet, Marston Bigot, Frome, Somerset, BA11 5BP, UK

Contact: Mrs M Chapman

National Motorcycle Council
Cowley House, 9 Little College Street, London, SW1E 3XS, UK

Tel: 071 222 0664
Fax: 071 233 0335

National Scooter Riders Association
PO Box 32, Mansfield, Nottinghamshire, NG19 0AZ, UK

Contact: Jeff Smith
Tel: 0623 651658

The N.S.R.A is the UK's largest organisation for scooter enthusiasts and organises national and international rallies, sporting weekends and various social events for its members along with their own prestigious annual scooter trade and custom show. Membership is annual (Jan-Jan) and gives members access to the N.S.R.A. insurance scheme and dealerwise, which are both discount schemes. The N.S.R.A. is BMF affiliated.

National Scooter Riders Association
34 Coronation Road, Forest Town, Mansfield, Nottinghamshire, NG19 0AJ, UK

Contact: Jeff Smith

National Sprint Association
66 Henconner Lane, Bramley, Leeds, Yorkshire (W.), UK

Contact: Roy Fisher

Never Too Tired: National Trike Owners
Club, 8 Lower Street, Macclesfield, Cheshire, SK11 7NJ, UK

New Era MCC
107 Mill Studio Business, Centre, Crane Mead, Ware, Hertfordshire, SE12 9PY, UK

Contact: Jean Maslin
Tel: 0920 444205
Fax: 0920 468686

The largest and most popular motor cycle road racing club in Britain. Organising some eighty meetings each year on all the major circuits.

New Era MCC
6 St Peters Avenue, Rushden, Northamptonshire, NN10 9XW, UK

Contact: Mrs J Mobley

New Imperial Owners Association
3 Fairview Drive, Higham, Kent, ME3 7BG, UK

Contact: Mike Slater

Norfolk & Suffolk Group
33 Falcon Road East, Norwich, Norfolk, NR7 8XZ, UK

Contact: Mrs M Armes
Tel: 0603 425749

North Devon British MC Owners Club
47 Old Town, Bideford, Devonshire, EX39 3BH, UK

Contact: Dai Davies
Tel: 0237 472237

Open to owners of any British bike. Meet 2nd & 4th Thursdays. Full calendar of runs, rallies, classic shows, green lanes. 1994 membership £7.00.

North East British Motorcycle Enthusiasts Club
9 Sawmill Estate, Alnwick, Northumberland, NE66 2QW, UK

Contact: J R Mark

Norton Owners Club
Beeches, Durley Brook Road, Durley, Southampton, Hampshire, SO3 2AR, UK

Contact: *Dave Fenner*
Tel: 0703 693262

Increase your enjoyment of owning a Norton with magazine, technical advice, library, spares, regalia, branches, rallies, shows and discounts with dealers, insurance and RAC.

Norton Racing Supporters Club
13 Belgrave Avenue, Saltney, Chester, Cheshire, CH4 8TY, UK

Contact: *Anne Collins*

Off Road Riders Scooter Association
53 Glenmore Drive, Woodshires, Coventry, Midlands (W.), CV6 6LZ, UK

Contact: *Brian Forde*

Ogri MCC
Tongrubben Wg 21/1, Enger, 4104, Germany

Contact: *Graham Kendall*

Oxford Speedway
Oxford Sports & Leisure, Sandy Lane, Cowey, Oxfordshire, OX4 5LJ, UK

Contact: *Tony Mole*
Tel: 0562 745683

Panther Owners Club
Willow Holt, Willow Hall Lane, Thorney, Peterborough, Cambridgeshire, PE6 OQN, UK

Contact: *J & A Jones*

Catering for over 600 devotees of the P+M marque. Hundreds of spares, workshop manuals, parts lists, technical information and social events.

Pat Watts Practice Track
114 St John Street, Bridgwater, Somerset, TA6 5HZ, UK
Tel: 0278 425098

Peterborough Speedway
East of England Showground, Alwalton, Peterborough, Cambridgeshire, PE2 OXE, UK

Contact: *Peter Oakes*
Tel: 0733 394494

Pickering & District Motor Club
4 Brockfield Road, Huntingdon Road, York, Yorkshire (N.), YO3 9DZ, UK

Contact: *Mr D A Brown*
Tel: 0904 622274

Pickering and District Motor Club promotes motorcycle grass track, motocross and trials. The club has been in existence since 1950.

Pontypool & DMCC
Freeholdland Post Office, Pontnewynydd, Pontypool, Gwent, UK

Contact: *Mrs T Bounds*
Tel: 0495 763096

Poole Speedway
Poole Stadium, Wimbourne Road, Poole, Dorset, BH15 2BP, UK

Contact: *Peter Ansell*
Tel: 0202 681145

Post Office Vehicle Club
7 Bignal Rand Drive, Wells, Somerset, BA5 2EU, UK

Contact: *John Targett*

Founded 1962 for those interested in post office and telephone vehicles past and present - monthly magazine, books, fleet lists, rallies, visits preservations and overseas coverage.

Pre-65 Motocross Club
Redesdale, Poyle Road, Tongham, Farnham, Surrey, GU10 1DS, UK

Contact: *Mrs N Ashley Barker*

Prison Motorcycle Brotherhood UK
PO Box 16, Corby, Northamptonshire, NN18 9NN, UK

Professional & Executive Motorcyclists Club
10 Broadhurst Drive, Kennington, Ashford, Kent, TN24 9RQ, UK

Contact: *Ed Gibson*

Putley Hill Farm Motocross Practice Track
Hill Farm, Putley, Hereford & Worcester, UK

Tel: 0531 670 552

RAC Motoring Services
PO Box 700, Spectrum, Bond Street, Bristol, Avon, BS99 1RB, UK

Racing 200 Club
9 Janefield Place, High Blantyre, Strathclyde, UK

Contact: *W Robertson*
Tel: 0698 824397

RAF Motor Sports Association
RAF Brampton, Huntingdon, Cambridgeshire, PE18 8QL, UK

Contact: *J Jones*

Raleigh Safety 7 & Early Reliant Owners Club
17 Courtland Avenue, Chingford, London, E4 6DU, UK

Contact: *Mick Sleap*

Reading Speedway
Bennett Road, Smallmead, Reading, Berkshire, RG2 OJL, UK

Contact: *Pat Bliss*
Tel: 0793 762722

Rhins Motorcycle Club
9 Bayview Road, Stranraer, Dumfries & Galloway, UK

Contact: *Alan McColm*
Tel: 0776 2829

Rickman Enfield Register
44 Maiden Erleigh Avenue, Bexley, Kent, DA5 3PE, UK

Contact: *Ian Abrahams*

Rickman Owners Club
35 Otterbourne Road, Chingford, London, E4 6LL, UK

Contact: *Mick Foulds*

Riders Union
31a St Georges Road, Leyton, London, E10 5RH, UK

Rochdale Owners Club
57 West Avenue, Handsworth Wood, Birmingham, Midlands (W.), B20 2LT, UK

Contact: *Brian Tomlinson*

Rockingham Forest Tourism Association
Civic Centre, George Street, Corby, Northamptonshire, NN17 1QB, UK

Tel: 0536 402551
Fax: 0536 400200

Bags of information for the tourist - day visits/short stay/leisure drives etc.

Royal Enfield Owners Club
Meadow Lodge Farm, Henfield, Coalpit Heath, Avon, BS17 2UX, UK

Contact: *John Cherry*

The club promotes the interchange of ideas and information on the maintenance, restoration and use of Royal Enfield motor cycles and machines.

Royal Navy MCC
HMS Daedalus, Lee on Solent, Hampshire, UK

Contact: *D Moore*

Rudge Enthusiasts
195 Filton Avenue, Bristol, Avon, BS7 0AY, UK

Contact: *Tim Berry*

Rye House Speedway
Rye House Stadium, Rye Road, Hoddesdon, Hertfordshire, UK

Contact: *Ron Russell*
Tel: 0263 514532

Scott Owners Club
2 Whiteshott, Basildon,
Essex, SS16 5HF, UK

Contact: H.W.Beal

First class journal. Library
service, spares scheme, technical
advice. Register of Scott
motorcycles, rallies and events.
Regular meetings of sections
throughout U.K., Australia, New
Zealand and U.S.A.

**Scottish Auto-Cycle
Association**
Block 2, Unit 6, Whiteside
Ind. Estate, Bathgate,
Lothian, EH48 2RX, UK

Governing body of motor cycle
sport in Scotland.

**Scottish Borders Enduro
Club**
53 Howden Crescent,
Jedburgh, Borders,
TD8 6JY, UK

Contact: Ian Bruce
Tel: 0835 62343

**Scottish Classic Racing
MCC**
113 Brediland Road,
Linwood, Strathclyde,
PA3 3RS, UK

Contact: Agnes Cadger
Tel: 0505 29759

**Scottish Vintage Racing
Club**
32 Regent Place, Edinburgh,
Lothian, UK

Contact: Andrew Johnstone
Tel: 031 661 2290

Sheffield Speedway
Owlerton Sports Stadium,
Penistone Road, Sheffield,
Yorkshire (S.), S6 2DE, UK

Contact: Neil Machin
Tel: 0742 853142

Sidecar Register
3 Rose Close, Kempshott,
Basingstoke, Hampshire,
RG22 5NF, UK

Contact: Bill Nicol

South Birmingham MCC
14 Spring Grove Road,
Kidderminster, Hereford &
Worcester, UK

Contact: Mr T Fairbrother
Tel: 0562 754827

**Spartan Kawasaki
Owners Club**
131 Fairfield Crescent,
Edgware, Middlesex, HA8
9AL, UK

Contact: Lou Rossetti

Speedway Association
15 Queen Elizabeth Drive,
Carrington, Essex, SS17
7TH, UK

**Speedway Riders
Association**
8 Balmoral Avenue,
Cheadle Hume, Cheshire,
SK8 5EQ, UK

**St Neots British Bike
Club**
2 Alamein Court, Eaton
Socon, St Neots,
Cambridgeshire, UK

Starters Driving Centre
Hadrian's Camp, Brampton
Old Road, Carlisle,
Cumbria, UK

Contact: Mike Evans
Tel: 0228 515002

**Stevenson & District
MCC**
Hillberry, 40 Seafield Court,
Ardrossan, Strathclyde,
KA22 8NS, UK

Contact: Norman Lamont
Tel: 0294 66346

Streetbike Drag Club
17 Southampton Road,
London, NW5 4JS, UK

Contact: Mr T Huck

Suffolk Moto Parc
Hill Farm, Rushmore,
Ipswich, Suffolk, UK

Tel: 0473 659222

Sunbeam MCC
Bleak House, 72 Chart
Lane, Reigate, Surrey,
RH2 7EA, UK

Contact: David Jordan

**Sunbeam Owners
Fellowship**
Rotor, PO Box 7, Market
Harborough, Leicestershire,
LE16 8XL, UK

**Sussex British
Motorcycle Owners Club**
96 Westlands, Rustington,
Littlehampton, Sussex (W.),
BN17 6RT, UK

Contact: C.E.Betts
Tel: 0903 716348

Open to all owning or interested
in British bikes. Meet every
Monday night at climbing sports
& social club. Membership £7.00,
OAP £3.50 PA.

Suzuki Owners Club
46 Lorne Street, Burslem,
Stoke On Trent,
Staffordshire, ST6 1AR, UK

Contact: Phil Hingert

Swindon Speedway
Abbey Stadium, Blunsden,
Swindon, Wiltshire,
SN2 4ND, UK

Contact: Peter Ansell
Tel: 0202 681 145

TT Riders Association
21 Stringhams Copse,
Ripley, Surrey, GU23 6JE,
UK

A registered charitable and
sociable association of
competitors who have raced in
the Isle of Man T.T. Claims to be
"the most exclusive club".

TT Supporters Club
50 Lyndhurst Road,
Birmingham, Midlands (W.),
B24 8QS, UK

Contact: Roy Hanks

**Tamworth & District
Classic MCC**
2 Meadow Park, Lichfield
Road, Tamworth,
Staffordshire, B79 7RR, UK

Contact: R K Salmon

Thruxton Moto Parc
10 North Way, Thatcham,
Berkshire, RG13 4BX, UK

Tel: 0635 865493

**Trail Bike Fun Enduro
Club**
80 Coledale Drive,
Stanmore, Middlesex,
BA7 2QF, UK

Contact: Don Felgate

Trail Riders Fellowship
Glebe House, St.Columb
Minor, Newquay,
Cornwall, TR7 3HD, UK

Trail Riders Fellowship
18 Corsham Road, Fords
Farm, Calcott, Reading,
Berkshire, RG3 5ZH, UK

Contact: Keith Forthergill

**Trident & Rocket 3
Owners Club**
50 Sylmond Gardens,
Rushden,
Northamptonshire,
NN10 9EJ, UK

Contact: H.J.Allen
Tel: 0933 317465

For Triumph and BSA triples pre
1976; offering bi-monthly
magazine, specialised technical
advice, D.V.L.A. machine dating,
archive material, machine
register, rallies, Beezumph Track
day.

Trike Quad Racing Club
5 Queensway, Delting,
Maidstone, Kent,
ME14 4LA, UK

Contact: Miss J Latter

Triumph Owners MCC
101 Great Knightleys,
Basildon, Essex, SS15 5AN,
UK

Contact: Mrs E Page

Two Wheel Centre & Youth Club
Chapman Street, Trail Park, Kingston upon Hull, Humberside (N.), HU8 7BU, UK

Tel: 0482 210676

Ural Riders Association
2 Cody Road, Clapham, Bedfordshire, MK41 6ED, UK

Contact: Deanna Dougan

VMC Highland Section
Morven, 42a Broadstone Park, Inverness, Highland, IV2 3LA, UK

Contact: A Alexander

VMC North East Section
6 Middleton Drive, Bridge Of Don, Aberdeen, Grampian, AB22 8LE, UK

Contact: J Addison

VMC Stirling Castle Section
10 Miller Place, Airth, Falkirk, Stirlingshire, FK12 8JY, UK

Contact: P Diamond
Tel: 0324 831846

Vegan Bikers Association
48 Hawkins Hall Lane, Datchworth, Knebworth, Hertfordshire, SG3 6TE, UK

Contact: J Coleman

Velocette Owners Club
1 Mayfair, Tilehurst, Reading, Berkshire, RG3 4RA, UK

Contact: V.Blackman

Social club with 27 centres in UK and overseas for those interested in Velocettes. Club has own spares company catering mainly for older models.

Veteran Speedway Riders Association
98 Colway Road, Penn, Wolverhampton, Midlands (W.), WV3 7NB, UK

Contact: Ron Hoare

Veteran Vespa Club
9A Coronation Road, Prestbury, Cheltenham, Gloucestershire, GL52 3DA, UK

Tel: 0242 244011

Now in its 40th year, the V.V.C provides a quarterly newsletter, spares service, annual rallies, books, manuals and road tests to help and encourage fellow owners in preserving and riding their machines. Send SAE for a membership form.

Vincent HRD Owners Club
Ashley Cottage, 133 Bath Road, Atworth, Wiltshire, SN12 8LA, UK

Contact: Information Officer

The V.O.C. for enthusiasts of HRD and Vincent motorcycles has world wide membership, a full diary of social and sporting events and a spares company.

Vincent HRD Series A Owners Club
74 Stonehill Avenue, Birstall, Leicester, Leicestershire, LE4 4JB, UK

Contact: B Stafford

Vincent HRD Series D Owners Club
18 Crabtree Road, Camberley, Surrey, GU15 2SY, UK

Contact: D Hills

Vintage Japanese MCC
24 North Road, Hordean, Portsmouth, Hampshire, PO8 OEH, UK

Contact: Maureen Head

Vintage Motor Cycle Club Ltd
Allen House, Wetmore Road, Burton on Trent, Staffordshire, DE14 1SN, UK

Tel: 0283 40557
Fax: 0283 510547

Vintage Motor Scooter Club
11 Ivanhoe Avenue, Lowton St Lukes, Warrington, Cheshire, WA3 2HX, UK

Contact: Ian Harrop

Vintage MCC Ltd
Central Scottish Section, 41 Coltward, Campmuir, Coupar Angus, Perthshire, PH13 9LH, UK

Contact: Mr A Burt

Walker Wheels
Riverside Facility, Pottery Bank, Walker, Newcastle upon Tyne, Tyne & Wear, NE6 3SX, UK

Contact: Harry Rogers
Tel: 091 265 5116

Warden Law Off-Road Centre
Sunderland Leisure, Mautland Square, Houghton le Spring, Tyne & Wear, DH4 4BL, UK

Tel: 091 512 0444

Watton DMC & CC
76 Bush Road, Hellesdon, Norwich, Norfolk, NR6 7UE, UK

Contact: Mr B Catchpole
Tel: 0603 402541

Westhoughton Classic & Modern MCC
16 Pendennis Avenue, Lostock, Bolton, Manchester, Gt., BL6 4RS, UK

Contact: Helen Tress

Westland Motorcycle Club
74 Preston Grove, Yeovil, Somerset, UK

Contact: Pat Brooks

Wolverhampton Speedway
Ladbroke Stadium, Sutherland Ave., Monmore Green, Wolverhampton, Midlands (W.), WV2 2JJ, UK

Contact: Chris Van Straaten
Tel: 021 520 2376

Women In The Wind
PO Box 143, Newbury, Berkshire, RG13 3HE, UK

Women's International Motorcycle Assoc.
2 Church Grove, Ladywell, London, SE13 7UU, UK

Contact:
Sheonagh Ravensdale
Tel: 081 690 4375

Women's International Motorcycle Assoc.
The Stables, 27 Hassocks Road, Hurstpierpoint, Sussex, UK

Contact: Vanessa Bretton

Worldwide Norton Riders
9 Sandon Close, Greswell, Stoke on Trent, Staffordshire, UK

Contact: John Clarke

Yamaha Owners Club
Fonzie, 72 Bookerhill Road, High Wycombe, Buckinghamshire, HP12 4EX, UK

Yamaha Riders Association
38 Frederick House, Pett Street, Woolwich, London, SE18 5PB, UK

Contact: Paul Watts

Youth Motorcycle Club (Scot) Ltd
Langton Lees, Duns, Berwickshire, UK

Contact: Kath MacVie
Tel: 0361 83220

Youth Motorcycle Sport Association
20 Kensington Gardens, Ilkeston, Derbyshire, UK

DEALERS

A & D Motorcycles
Spencer Trading Estate,
Colomendy, Denbigh,
Clwyd, LL16 5TQ, UK

Tel: 0745 815105

**A & D Verity
Motorcycles**
7 East Street, St Ives,
Huntingdon,
Cambridgeshire, PE17 4PB,
UK

Tel: 0480 63637

APS Motorcycles
12-18 Stokes Craft, Bristol,
Avon, BS1 3PR, UK

Tel: 0272 423602

A44 Motorcycles
3 Etnam Street, Leominster,
Hereford & Worcester, UK

Tel: 0568 612564

Large selection of quality used
machines; P/X & H/P terms;
bikes bought for cash; helmets
and clothing; computerised parts
service; Yokohama tyres, etc.

ACS International
2277 West Highway, St
Paul, Minnesota, 55113,
USA

Tel: 612 633 9655

Abingdon Motorcycles
Marcham Rd., Abingdon,
Oxon, OX14 1TZ, UK

Tel: 0235 550055

Alford Bros (Folkestone)
20 Cheriton Road,
Folkestone, Kent,
CT20 1BU, UK

Tel: 0303 254057

All States Motorcycles
229 London Road, Reading,
Berkshire, RG1 3NY, UK

Tel: 0734 352804

Alvins Motorcycles
9B Springfield Street,
Edinburgh, Lothian,
EH6 5EF, Scotland

Tel: 031 554 4155
Fax: 031 553 3093

Main dealers for
Harley-Davidson, Suzuki, and
Ducati; full sales, spares and
service facilities. Also Scotland's
largest and best stocked clothing
and helmet centre.

**American & Japanese
Classic Export**
RR2 Box 23, Route 45
South, Mattoon, Illinois,
61938, USA

Tel: 0101 217 235 2934

Anchor Kawasaki Centre
99 Wellingborough Road,
Finedon,
Northamptonshire,
NN9 5OG, UK

Tel: 0933 680274

**Andy Tiernan
Classics Ltd**
Old Railway Works, Station
Road, Framlingham,
Suffolk, IP13 9EE, UK

Tel: 0728 724321

Aprilia Moto UK Ltd
Unit 11, Gregory Way,
Reddish, Stockport,
Manchester, Gt., ST5 7ST,
UK

Tel: 061 476 5770

Ash Vale Motorcycles
6 Vale Road, Ash Vale,
Aldershot, Hampshire,
GU12 5JH, UK

Tel: 0252 312998

Ashley Banks Ltd
5 King Street, Langtoft,
Peterborough,
Cambridgeshire, PE6 9NF,
UK

Tel: 0778 560651

Atlantic Motor Cycles
20 Station Road, Twyford,
Berkshire, RG10 9NT, UK

Tel: 0734 342266

Aye Gee Motorcycles
211/219 Bellegrove Parade,
Welling, Kent, DA16 3RQ,
UK

Tel: 081 856 4194

B & M Imports Ltd.
Great Hanover Street,
Preston, Lancashire,
PR1 1PE, UK

Tel: 0772 59063/555964

B Wybrow M/C Ltd
26 Corringham Rd,
Stanford le Hope, Essex,
SS17 0AH, UK

Tel: 0375 672823
Fax: 0375 672823

Main Honda dealer closed Wed
and Sun. Clothing, helmets,
accessories, spares, servicing,
MOTs, autocom and Datatool
stockist. Just 10 minutes from
Junction 30, M25.

BMW (GB) Ltd
Ellesfield Avenue, Bracknell,
Berkshire, RG12 4TA, UK

Tel: 0344 426565

BVM Moto
16 Slad Road, Stroud,
Gloucestershire,
GL5 1QW, UK

Tel: 0453 762743
Fax: 0453 753972

B.M.W. motorcycle specialists,
many used Japanese road and
off road machines. Insurance,
parts, clothing, service.

Bat Motorcycles
30 High Street, South
Norwood, London,
SE25 9HA, UK

Tel: 081 653 2347

Baycover Ltd
25A Whapload Road,
Lowestoft, Suffolk,
NR32 1UH, UK

Tel: 0502 582352

Best Of Bikes
4 Oxford Street, Newbury,
Berkshire, RG13 1JB, UK

Tel: 0635 30004

Better Bikes Ltd
1-2 Richmond Lane,
Edinburgh, Lothian,
EH8 9HS, UK

Tel: 031 667 9177

Bike Business
47-49 Fratton Road,
Portsmouth, Hampshire,
PO1 5AB, UK

Tel: 0705 832575

Bike Spurz
27 Junction Lane, Sutton,
St Helens, Merseyside,
WA9 3JN, UK

Bike Treads
Unit 7, Ash Kembrey Park,
Swindon, Wiltshire,
SN2 6UN, UK

Tel: 0793 615995
Fax: 0793 615995

Stockists of tyres, chains,
sprockets, brakes and exhausts,
etc. Full fitting service while you
wait. Mail order by
Access/Visa/Switch. Open
Monday to Saturday.

Bikerama Ltd
42-50 High Street,
Hornsey, London,
N8 7NX, UK

Tel: 081 348 1771

Bikesport
208 Westgate Road,
Newcastle upon Tyne, Tyne
& Wear, NE4 6AN, UK

Contact: G R Linscott
Tel: 091 232 8970

Bill Little Motorcycles
Oak Farm, Braydon,
Swindon, Wiltshire,
SN5 0AG, UK

Tel: 0666 860577

Bill Pope (Motors) Ltd
51-59 Winwick Road,
Warrington, Cheshire,
WA2 7DG, UK

Tel: 0925 34131

Bill Roberts Race Fittings
Green Acres, Cranfield
Park Avenue, Wickford,
Essex, UK

Tel: 0268 726571

Bill Smith Motors
Moor Place, Liverpool,
Merseyside, L3 5QW, UK

Tel: 051 709 0677

Bimota Kinetic Art Ltd
Loxwood Village,
Billingshurst, Sussex (W.),
RH14 6QS, UK

Tel: 0403 752644

Birmingham Motorcycles Ltd
1163-71 Pershore Road,
Stirchley, Birmingham,
Midlands (W.), B30 2YJ, UK

Tel: 0214 725088

Bits & Pieces
45 Baddow Road,
Chelmsford, Essex, UK

Tel: 0245 351196

Bob Hill Motorcycles
7 Cowbridge, Hertford,
Hertfordshire, SG14 1PG,
UK

Tel: 0992 551711

Bob Jackson
Rose & Crown Works,
Stricklandgate, Kendal,
Cumbria, UK

Tel: 0539 720582

Bollenbach Engineering
296 Williams Place, East
Dundee, Illinois,
60118-2319, USA

Tel: 708 428 2800

Boyer
151 Plumstead Road,
London, SE18 7DY, UK

Tel: 081 854-8133
Fax: 081 855-5130

Triumph & Kawasaki motor cycle sales, spares, service and MOTs. 9.00-6.00 Mon., Tue., Wed. and Fri.; closed Thurs.; Sat. 9.00-5.00.

Brentwood Cycles
5 Crown Street,
Brentwood, Essex,
CM14 4BA, UK

Tel: 0277 212423

Brian Gray Motorcycles
Station Road, High
Wycombe,
Buckinghamshire,
HP13 6AD, UK

Tel: 0494 438615

Brian Lee Motorcycles
40 Mary Street,
Scunthorpe, Humberside
(S.), DN15 6NR, UK

Tel: 0724 844069

Bridge Motorcycle World
1-3 Verney Street, Exeter,
Devonshire, EX1 2AW, UK

Tel: 0392 432654

British Iron Company
Rossley, Snowhill Road,
Crawley Down, Sussex
(W.), RH10 3HA, UK

Tel: 0342 717461
Fax: 0342 717611

Bygone Motorcycles
348 Manchester Road,
Warrington, Cheshire,
WA1 3RE, UK

Tel: 0925 831477

Bykebitz Motorcycle Parts
Reading Road, Yateley,
Camberley, Surrey,
GU17 7UR, UK

Tel: 0252 870900
Fax: 0252 877079

Largest stock of parts for Japanese machines in the area. Clothing and accessories, importers for Clymer, Sola-larm, Snug products. Open Mon-Sat.

C J Wilson
23-25 West Main Street,
Uphall, Broxburn, Lothian,
UK

Tel: 0506 856751

C.G. Chell Motorcycles
25-27 Marston Road,
Stafford, Staffordshire,
ST16 3BS, UK

Tel: 0785 42356

C.J. Bowers Ltd
Ribsygate Street, Bury St
Edmunds, Suffolk,
IP33 3AD, UK

Tel: 0284 705726

C.M. Smith
87 Worthington Road,
Fradley, Lichfield,
Staffordshire, UK

Tel: 0543 263922

CBS (Whitton) Ltd
136 Kneller Road,
Twickenham, Middlesex,
TW2 7DX, UK

Tel: 081 898 5492
Fax: 081 894 3292

Kawasaki and scooter specialist. Spares and repairs for Kawasaki, Jawa, Tomos, Vespa and Piaggio scooters. Finance and insurance arranged. Closed Mondays.

CMW Motorcycles Ltd
20 The Hornet, Chichester,
Sussex (W.), PO19 4JG, UK

Tel: 0243 782544
Fax: 0243 532761

Yamaha, Ducati, Moto Guzzi and Cagiva dealers. Full workshop facilities, spares, accessories, clothing, Furygan leathers. Used motorcycles bought for cash. Closed all day Thursdays.

CW Motorcycles
Gt. Western Ind. Ctr.,
Dorchester, Dorset, DT1
1QW, UK

Tel: 0305 269370

Cagiva Three Cross (Imports) Ltd
Woolsbridge Ind.Est.,
Old Barn Farm Road,
Three Legged Cross,
Wimbourne, Dorset, UK

Tel: 0202 823344

Callington Kawasaki
Unit 7, Mosside Industrial
Estate, Callington,
Cornwall, UK

Tel: 0579 82286

Caravan Camping & Leisure
42 Cromer Road, West Runton, Norfolk, NR27 9AD, UK
Tel: 0263 837482

Camping and outdoor (mail order) * tents * sleeping bags * mats * stoves * lights * cookware * accessories * spares * repair services * plus much more! Free mail order lists.

Carl Rosner Ltd
Station Approach, Sanderstead Road, South Croydon, Surrey, CR2 OPL, UK
Contact: Carl Rosner
Tel: 081 657 0121
Fax: 081 651 0596

Established 25 years. Comprehensive spares stock for Commando, Triumph 650/750. Trident/R-3. Full workshop facilities, including restorations. Dealers for Hinckley Triumphs, Norton Rotary and Indian Enfield.

Carnell Motor Group
184-188 Beverley Road, Hull, Humberside (N.), HU3 IVS, UK
Tel: 0482 28573

Carnells
16/17 Autocentre, Stacey Bushes, Milton Keynes, Buckinghamshire, MK12 6HS, UK

Tel: 0908 221802
Fax: 0908 310806

Complete dealer. Open 7 days. Suzuki, Honda, Kawasaki. Huge stock quality used motorcycles. Extensive range of clothing, many discounted items. Workshop. Discount tyres. Insurance.

Carnells
Marshgate, Doncaster, Yorkshire (S.), DN5 8AF, UK

Tel: 0302 321383
Fax: 0302 344041
Complete dealer. Open 7 days. All Japanese franchises. Over 200 quality used motorcycles. Extensive range of clothing, many discounted items. Workshop. Discount tyres. Insurance.

Channel Islands St Peter Port Garage
Trinity Square, St Peter Port, Guernsey, CI, UK

Tel: 0481 724261

Claremont Motorcycles (Gloucester)
Claremont Road, Gloucester, Gloucestershire, GL1 3NY, UK

Tel: 0452 525903
Fax: 0452 306475

Honda and new Triumph dealer plus full parts servicing, MOT and clothing department.

Clarks Motorcycles
63 Osmaston Road, Derby, Derbyshire, DE1 2JH, UK

Tel: 0332 293760

Classic Bikes
17 Church Road, Redfield, Bristol, Avon, UK

Tel: 0272 557762
Fax: 0272 606102

Classics In Cheshire
P.O.Box 1, Northwich, Cheshire, CW8 2RD, UK

Tel: 0606 888888

Clay Cross Kawasaki
Broadley, Clay Cross, Chesterfield, Derbyshire, S45 9JP, UK

Tel: 0246 260124

Clive Castledine Motor Cycles
141-143 Humberstone Road, Leicester, Leicestershire, LE5 3AP, UK

Tel: 0533 620966

Colin Appleyard Ltd
Wellington Road, Worth Way, Keighley, Yorkshire (W.), BD21 5AJ, UK

Tel: 0535 606311

Colin Collins Motorcycles
26-28 High Street, Wealdstone, Middlesex, HA3 7AR, UK

Tel: 081 861 1666

Colin Lomax Motorcycles
3 Ray Street, Heanor, Derbyshire, DE7 7GE, UK

Tel: 0773 713475

Colwyn Bay Motorcycles
309 Abergele Road, Old Colwyn, Colwyn Bay, Clwyd, LL29 9YF, UK

Corby Kawasaki Centre
Phoenix Parkway, Corby, Northamptonshire, NN17 IDY, UK

Tel: 0536 401010

Cradley Heath Kawasaki
St Annes Road, Cradley Heath, Midlands (W.), B64 5BJ, UK

Tel: 0384 633455

Craigie Competition Motorcycles
52 Crookstown Industrial Est., Tallaght, Dublin, Eire

Tel: 010 3531 525866

Crescent Suzuki Bournemouth
324/6 Charminster Road, Bournemouth, Dorset, UK

Tel: 0202 512923

Crooks Suzuki Ltd
36-44 Crellin Street, Barrow in Furness, Cumbria, LA14 IDY, UK

Tel: 0229 822342

Cross Motorcycles
42 New Street, St.Helier, Jersey, CI, UK

Tel: 0534 33911

Cross Street Kawasaki
The Square, Cross Street, Enderby, Leicestershire, LE9 5NJ, UK
Tel: 0533 866147

Crowmarsh Classic Motorcycles
Unit 3, Honey Farm, Preston, Crowmarsh, Oxfordshire, OX10 6SL, UK

Tel: 0491 25869

Cusworths Kawasaki
8 Wood Street, Doncaster, Yorkshire (S.), UK

Tel: 0302 323703

Dave Cooper
88 Mount Culver Avenue, Sidcup, Kent, DA14 5JW, UK

Tel: 086 070 2112

Dave Fox Motorcycles
146-148 King Street, Ramsgate, Kent, CT11 8PJ, UK

Tel: 0843 591113

David Jones Motorcycles
Poole Road, Newton, Powys, SY16 IDD, UK

Tel: 0686 625010

Daytona
42-48 Windmill Hill, Ruislip Manor, Middlesex, HA4 8PT, UK

Tel: 0895 675511
Fax: 0895 630654

Dealership for Kawasaki, Triumph and Ducati motorcycles. Also main parts stockist for Kawasaki.

Denver Motorcycles
1 Sluice Road, Downham Market, Norfolk, UK

Tel: (0366) 382115

Des Helyar Motorcycles
The Forge Garage, Pamber Heath Road, Tadley, Hampshire, UK

Tel: 0734 700665

Desperate Dans
17a-17b Old Road, Linslade, Leighton Buzzard, Bedfordshire, LU7 7RB, UK

Direct Tyres
Marshgate, Doncaster,
Yorkshire (S.), DN5 8AF,
UK

Tel: 0302 368466
Fax: 0302 344041

Select from our huge range of
top tyres from all the major
marques, including many with
excellent discount. Try our mail
order hotline.

**Dirtwheels of Coventry
Ltd**
London Road, Ryton on
Dunsmore, Coventry,
Midlands (W.), CV8 3EW,
UK

Tel: 0203 301852

Doble Motorcycles
86-88 Brighton Road,
Coulsdon, Surrey,
CR5 2NA, UK

Contact: Mike Doble
Tel: 081 668 8851
Fax: 081 668 6100

Honda specialist dealer. Full
range of new Hondas in stock
and over 100 used machines.
Servicing, MOTs, insurance,
crash estimates, part exchange
and finance arranged. Open six
days a week from 8.30 - 6.30,
situated 2 1/2 miles off the M25
on the A23 in Coulsdon.

**Doug Hacking
Motorcycles**
141-148 Chorley Old Road,
Bolton, Manchester, Gt.,
UK

Tel: 0204 491511

Drayton Croft M/C
323-325 Foleshill Road,
Coventry, Midlands (W.),
CV1 4JS, UK

Tel: 0203 665433

**E A Grimstead & Sons
Ltd**
261-265 Barking Road, East
Ham, London, E6 1LB, UK

Tel: 081 552 9912

E.C. Bate
60/62 West Hill, Dartford,
Kent, DA1 2EU, UK

Tel: 0322 220748

Earnshaws Ltd
Manchester Road,
Huddersfield, Yorkshire
(W.), HD1 3LE, UK

Tel: 0484 421232

East End Motorcycles
224 Newtonards Road,
Belfast, Co. Antrim, BT4
1HB, N.Ireland

Tel: 0232 731454

Eddy Grimstead Ltd
741-755 Eastern Avenue,
Newbury Park, Ilford,
Essex, IG2 7RT, UK

Tel: 081 590 6615

Open 9 to 6 Monday to Saturday.
A family business established in
1908. Main agents for Honda,
Suzuki, Yamaha & Kawasaki.
Located A12 near Newbury Park
Station.

Elbe Motocross
524-532 Stoney Stanton
Road, Coventry, Midlands
(W.), UK

Tel: 0203 687049

**Enfield Bavanar
Products Ltd**
47 Beaumont Road, Purley,
Surrey, CR8 2EJ, UK

Tel: 081 645 9190

Ernie Page Motors Ltd
42 Polworth Crescent,
Edinburgh, Lothian, UK

Tel: 031 229 5655

F H Warr & Sons
104 Waterford Road,
London, SW6 2EU, UK

Tel: 071 736 2934

Fair Spares
The Corner Garage,
96-98 Cannock Road,
Burntwood, Staffordshire,
WS7 8JP, UK

Tel: 0543 278008
Fax: 0543 274775

Retailers in Norton and Triumph
Twin spares, repairs and
renovations. Norton
Commandos built new to order.

Felbridge Garage
Felbridge, East Grinstead,
Sussex (W.), RH19 2RH,
UK

Tel: 0342 324677

Flitwick Motorcycles
Station Road, Flitwick,
Bedfordshire, UK

Tel: 0525 712197

Ford & Ellis Motorcycles
132-152 Broad Street,
Chesham, Buckinghamshire,
HP5 3ED, UK

Tel: 0494 772343

Fowlers Of Bristol Ltd
2-12 Bath Road, Pylie Hill,
Bristol, Avon, BS4 3DR,
UK

Tel: 0272 770466

Fox Motorcycles
326-330 Hucknall Road,
Sherwood,
Nottinghamshire, NG5 1FS,
UK

Tel: 0602 691462

Fox's of Worksop
4 Carlton Road, Worksop,
Nottinghamshire, SA10
1PM, UK

Tel: 0909 482614

**Frettons of Leamington
Ltd**
15-17 Clemens St,
Leamington Spa,
Warwickshire, UK

Tel: 0926 429214

**Frontiers Motorcycles
Ltd**
363 Kingston Road,
Wimbledon, London,
SW20, UK

Tel: 081 540 7774

GB Motorcycles
Withycombe, Station Road,
Christian Malford,
Wiltshire, SN15 4BG, UK

Tel: 0249 720448

GS Motorcycles
23 Lisburn Road,
Hillsborough, Co.Down,
BT26 6AA, UK

Tel: 0846 689777

GT Motorcycles
152-158 Albert Road,
Devonport, Plymouth,
Devonshire, PL2 1AG, UK

Tel: 0752 559063

Galleria Bimota Ltd
Stocklund Square, High
Street, Cranleigh, Surrey,
GU6 8RG, UK

Tel: 0483 272707

George Petch Wheels
133 Corporation Road,
Grimsby, Humberside,
DN31 1UR, UK

Tel: 0472 354402

**George White
Motorcycles**
7-8 Manchester Road,
Swindon, Wiltshire,
SN1 2AB, UK

Tel: 0793 522786

Gilera - Bob Wright MCs
4 Orchard Street, Weston
super Mare, Avon, UK

Tel: 0934 413847

Gloucester Kawasaki Centre
99-101/124-126 Barton Street, Gloucester, Gloucestershire, GL1 4DZ, UK

Tel: 525128
Fax: 309191

Massive stocks of motorcycles. Fast spares ordering for all Japanese makes. Repairs, servicing, large clothing and accessories department. Open six days a week 9am-6pm.

Gloucester Kawasaki Centre
124-126 Barton Road, Gloucester, Gloucestershire, UK

Tel: (0452) 306485/424190

Gloucestershire AMS Motorcycles
30 St James Street, Cheltenham, Gloucestershire, UK

Tel: 0242 583985

Goodridge (UK) Ltd
Exeter Airport, Exeter, Devonshire, EX5 2UP, UK

Tel: 0392 366956

Granby Motors
310-314 Radford Road, Nottingham, Nottinghamshire, NG7 5GN, UK

Tel: 0602 787077

Granby Motors (Ilkeston) Ltd
2-4 Pelham Street, Ilkeston, Derbyshire, DE7 8AR, UK

Tel: 0602 324961

Greens of Longton
Bridgewood Street, Longton, Stoke on Trent, Staffordshire, ST3 1HW, UK

Tel: 0782 312023

Hallens Motorcycles Ltd
Hawthorne Way, Chesterton Road, Cambridge, Cambridge-shire, CB4 1BG, UK

Tel: 0223 356225

Hamiltons Motorcycle Centre
420-446 Streatham High Road, London, SW16 3PX, UK

Tel: 081 764 0101
Fax: 081 764 4401

Motorcycle dealers for over fifty years. Main agents for Yamaha, Honda, Kawasaki and Suzuki. Large range of used machines. Computerised spares department. Workshop/MOT station.

Harley-Davidson UK Ltd
PO Box 27, Daventry, Northamptonshire, NN11 5RW, UK

Tel: 0327 301340

Harley-Davidson of Southport
153 Eastbank Street, Southport, Merseyside, PR8 1EE, UK

Tel: 0704 543745

Hawkshaw Motorcycles Ltd
37 Bridge Road, Blundellsands, Liverpool, Merseyside, L23 68A, UK

Tel: 051 924 2369

Heritage Harley-Davidson
1339A London Road, Leigh on Sea, Essex, SS9 2AB, UK

Tel: 0702 75883

Heron Suzuki Plc
P.O.Box 56, Tunbridge Wells, Kent, TN1 2XY, UK

Heron Suzuki Plc
Gatwick Road, Crawley, Sussex (W.), RH10 2XF, UK

Tel: 0543 480101

Holloway's - Auctioneers
49 Parsons Street, Banbury, Oxfordshire, OX16 8PF, UK

Tel: 0295 253197
Fax: 0295 252672

Auctioneers of collector's cars, motorcycles & automobilia. Enquiries from sellers and buyers invited.

Honda UK Ltd
Power Road, Chiswick, London, W4 5YT, UK

Tel: 081 747 1400

Hongdu - Anglo Hazet German Tools Ltd
Thrupp, Stroud, Gloucestershire, UK

Tel: 0453 886258

Hub Motor Cycles
229 Stapleton Road, Easton, Bristol, Avon, UK

Tel: 0272 510717

Ian Melrose
18A St Lukes Mews, London, W11 1DS, UK

Tel: 071 951 9139

International Motorcycles
Otto Hahn Strasse 7, Paderborn, Sennelagar, 4790, Germany

Tel: 010 49 5254 7723

Ireland Dublin Harley-Davidson
24-25 Blessington Street, Dublin 7, Eire

Tel: 010 3531 303000

Isle of Wight Motorcycles
7-9 Hunny Hill, Newport, Isle of Wight, PO30 5HJ, UK

Tel: 0983 522675

Italia Classics
Margram Filling Station, Monks Road, Lincoln, Lincolnshire, LN5 2PP, UK

Tel: 0522 511851

Italsport
Unit 1, Yarwood Street, off Rochdale Road, Bury, Manchester, Gt., BL9 7AU, UK

Tel: 061 797 6124

J Morgan
Brookfield Mill, Crumlin Road, Belfast, Co.Antrim, BT14 7EA, UK

Tel: 0232 757720

JT's Motorcycles
6 Castle Square, Swansea, Glamorgan (W.), SA1 DW, UK

Tel: 0792 461776

Jack Horseman Motorcycles
39 London Road, Carlisle, Cumbria, UK

Tel: 0228 45333

Jack Lilley Ltd
109-113 High Street, Shepperton, Middlesex, TW17 9BL, UK

Contact: S.M. Lilley
Tel: 0932 246055
Fax: 0932 244673

European bike centre, Triumph, Ducati, Moto Guzzi, Piaggio, MZ, Cagiva. Open Monday-Saturday, 9.00am-6.00pm, 5 mins J11 M25 or J1 M3. B.Rail Waterloo 40 mins.

Jack Nice Motorcycles
129-133 Grove Road, Walthamstow, London, E17 9BU, UK

Tel: 081 520 1920

Jacksons Motorcycles
22-23 Borough Road, Burton on Trent, Staffordshire, DE14 2DA, UK

Tel: 0283 65 154

Jawa - Skoda (GB) Ltd
Bergen Way, North Lynn Ind.Estate, King's Lynn, Norfolk, UK

Tel: 0553 761176

Jim Aim Motorcycles
147 Swan Street, Sible
Headingham, Halstead,
Essex, CO9 3PT, UK

Tel: 0787 60671

John E Vines Ltd
39 Elm Road, Leigh on Sea,
Essex, SS9 ISW, UK

Tel: 0702 72034

John Pease Motorcycles
41/43 Railway Street,
Braintree, Essex, CM7 6JD,
UK

Tel: 0376 321819

John W Groombridge
Mayfield Road Garage,
Cross in Hand, Heathfield,
Sussex (E.), UK

Tel: 0435 862466

Johns of Romford
46-52 London Road,
Romford, Essex,
RM17 9QX, UK

Tel: 0708 746293

K & M Motorcycles
18b Tower Road,
St Leonards on Sea, Sussex
(E.), UK

Tel: 0424 439767

K Southport Superbikes
86-90 Eastbank Street,
Southport, Merseyside, UK

Tel: 0704 536192

**Kawasaki (Paddock
Wood) Ltd**
62 Maidstone Road,
Paddock Wood, nr
Tonbridge, Kent,
TN12 6AF, UK

Tel: 0892 835353

Kawasaki Aberdeen
72 Hutcheon Street,
Aberdeen, Grampian,
AB2 3XE, UK

Tel: 0224 638894

Kawasaki Carlisle Ltd
30-32 Bridge Street,
Caldewgate, Carlisle,
Cumbria, CA2 5SX, UK

Tel: 0228 25024

**Kawasaki Distributors
Ltd**
17 Wood Street, Dublin 8, Eire
Tel: 01 756046

Kawasaki In London
151 Plumstead Road,
London, SE18 7DY, UK

Tel: 081 854 8133

Kawasaki Kingdom
96 London Road, Bexhill
On Sea, Bexhill on Sea,
Sussex (E.), TN39 3LE, UK

Kawasaki Motors Ltd
1 Dukes Meadow,
Millboard Road, Bourne
End, Buckinghamshire,
SL8 5XF, UK

Tel: 0628 851000

Kawasaki Newcastle
165 Westgate Road,
Newcastle upon Tyne, Tyne
& Wear, NE1 4AN, UK

Tel: 091 232 7010

Kawasaki Peterborough
723 Lincoln Road,
Peterborough,
Cambridgeshire, UK

Tel: 0733 341003

Ken's Motorcycles Ltd
195 & 246-255 Westgate
Road, Newcastle upon
Tyne, Tyne & Wear,
NE4 6AQ, UK

Tel: 091 232 1793

Kendall & Pitt
36A-42 Barrack Street,
Colchester, Essex,
CO1 2CR, UK

Tel: 0206 794849

Kevark Motorcycles
Victoria Street, Grimsby,
Humberside (S.),
DN31 1DJ, UK

Tel: 0472 357573

Kick Start Motorcycles
51 Talbot Road, Port
Talbot, Glamorgan (W.),
SA13 1HU, UK

Tel: 0639 881585

Knotts of Stratford
15 Western Road,
Stratford upon Avon,
Warwickshire, CV37 0AH,
UK

Tel: 0789 205149

L P Williams
Common Lane Industrial
Estate, Kenilworth,
Warwickshire, CV8 2EF,
UK

Contact: Trevor Gleadall
Tel: 0926 54948

L.W. Haggis
4 Peterhouse Parade,
Grattons Drive, Pound Hill,
Crawley, Sussex,
RH10 3BA, UK

Tel: 0293 886451

Lee Bros
41-47 Bolton Brow,
Sowerby Bridge, Yorkshire
(W.), HX6 2AL, UK

Tel: 0422 831727

Len Manchester
17 Burton Street, Melton
Mowbray, Leicestershire,
UK

Tel: 0664 62302

**Lexport (Sales &
Service) Ltd**
131-133 High Street, West
Wickham, London,
BR4 0LU, UK

Tel: 081 777 8040

M & S Weatherley
American Imports,
Harleston, Norfolk, UK

Tel: 0379 854067

MPS
Daneheath Business Park,
Heathfield, Newton Abbot,
Devonshire, TQ12 6TL, UK

Tel: 0626 835835
Fax: 0626 835152

The UK's largest mail order
motorcycle house, products
stocked include D.I.D. chain,
Vesram disc pads, Motad
exhausts, Goodridge brake lines,
Moto Fizz accessories, Abus
locks.

MZ Motorcycles GB Ltd
Unit 98-101, Northwick
Park Business Centre,
Blockley, Gloucestershire,
GL5 9RF, UK

Tel: 0386 700753

Marcol Motorcycles
17 Huntingdon Street,
Nottingham,
Nottinghamshire, NG1 3JH,
UK

Tel: 0602 507912

Nottingham's premier clothing,
spares, accessory and used
machine centre. Full range of
luggage, pattern parts and
security products. Machines
wanted, cash paid. P/ex and
finance offered.

**Marcol Motorcycles
(Leicester)**
23 Blackbird Road,
Leicester, Leicestershire,
LE4 0AH, UK

Tel: 0533 623554
Fax: 0533 623554

Leicester's premier clothing,
spares, accessory and used
machine centre. Full range of
luggage, pattern parts and
security products. Machines
wanted. Cash paid, part
exchange and finance.

Market Garage
Brook Street, Tring,
Hertfordshire, UK

Tel: 0442 822599

Medium Link Ltd
43-45 East Reach, Taunton,
Somerset, TA1 3ES, UK

Tel: 0823 272378

You've had it Soft for too Long.

All that comfort and character has made you complacent.

You've got used to the smooth power delivery, the frugal fuel economy, the endless torque and the mile eating comfort.

Now We've Made it Harder!

We've given the bike an edge. The power is increased by 15%, the brakes are stronger, the handling is improved and there is an incredible 26% more torque.

New silent block mountings for the footboards, hand grips and mirrors give you our smoothest ride ever. Changes include a new 1064cc engine, new frame, new suspension, choice of colours and a choice of luggage to give a greater custom image to the California. Altogether there are nearly 200 changes to the new Moto Guzzi California. Final prices vary according to specification, starting from only £6599 on the road. See your dealer now for a No Obligation Test Ride - We know we'll impress!

Changes include:

More Capacity to 1064 cc	Improved Brakes
New Design Pistons	New Design tail-light
Power Increased by 15%	An Even More Comfortable Seat
26% More Torque	
Fuel injection increased to 40mm bore	Completely Water-Tight Electrics
Carburettors from 30mm to 36mm	Bigger Tyre width
	A Choice of Luggage
New Camshafts	30 litre Rigid Bags
Bigger Airbox	40 litre Rigid Bags
New Anti-Corrosion Engine Paint	Rear Top Case
New, Smoother, Easy-Selection Gearbox	A Choice of Small or Large Windscreen
Stronger Frame	Leather Bags
	Leather Top Carrier
Suspension has More Travel and More Adjustment	Rear Chrome Guards

Pure Street

Soft Luggage

Full Touring

MOTO GUZZI
TOURING

Michael Freeman Motors
The Camp, Stroud.
Gloucestershire, GL6 7HN,
UK

Tel: 0285 821297

**Mickey Oates
Motorcycles**
19 North Canalbank Street,
Port Dundas Estate,
Glasgow, Strathclyde, UK

Tel: 0302 3327374

**Mid Wales Kawasaki
Centre**
Cross Gates, Llandrindod
Wells, Powys, UK.

Tel: 0597 851811

Mid-Devon Motorcycles
School Way, Market Street,
Okehampton, Devonshire,
EX20 1AN, UK

Tel: 0837 52517

Mike Bavin Motorcycles
101 Victoria Road, Diss,
Norfolk, IP22 3JG, UK

Tel: 0379 642631

Miles Kingsport Ltd
104 Whitham, Hull,
Humberside (N.),
HU9 1AT, UK

Tel: 0482 23529

Motex Ltd
Shire Business Park,
Warndon, Hereford &
Worcester, WR4 9FD, UK

Tel: 0905 756883

Moto's UK Ltd
18-20 Old Road, Linslade,
Leighton Buzzard,
Bedfordshire, LU7 7RE, UK

Tel: 0525 851818

**Motor Vehicle Imports
Ltd**
Western House, Middle
Lane, Wythal, Birmingham,
Midlands (W), B47 6LA, UK

Tel: 0564 824171

Motorcycle Centre
2-4 Carrington Road,
Stockport, Cheshire, SK1
2QE, UK

Tel: 061 480 3346

**Motorcycle Centre
(Hereford) Ltd**
Belmont Roundabout,
7-9 Ross Road, Hereford,
Hereford & Worcester,
HR2 7RH, UK

Tel: 0432 272341

Motorcycle City
30/32 Clapham High Street,
Clapham, London, SW4
7UR, UK

Tel: 071 720 6072

Motorcycle City Sales
470-478 Oxford Road,
Reading, Berkshire, G3
1EF, UK

Tel: 0734 574044

Motorcycle City Sales
55-59 Forton Road,
Gosport, Hampshire,
PO12 4TD, UK

Tel: 0705 581226

Motorcycle City Sales
300-304 Ruislip Road East,
Greenford, Middlesex,
UB6 9BH, UK

Tel: 081 578 3218

Motorcycle City Sales
537 Staines Road, Bedfont,
Middlesex, UK

Tel: 081 890 2913

**Motorcycle Clothing
Centre**
36 Norwich Road,
Wymondham, Norfolk,
NR18 0NS, UK

*Contact: Chris Clarke
Tel: 0953 606922*

Motorcycle Mart
22-25 Broadway, Roath,
Cardiff, Glamorgan (S.),
CF2 1QE, UK

Tel: 0222 484144

Motorcycle World
Talbot Green, Llantrisant,
Glamorgan, Mid-,
CF7 8AW, UK

*Contact: Mr M Williams
Tel: 0443 227903*

**Motorway Tyres &
Accessories Ltd**
Crown House, 10 Crown
Street, Reading, Berkshire,
RG1 2SL, UK

Neval Motorcycles
Brockholme, Seaton Road,
Hornsea, Humberside (N.),
HU18 1BZ, UK

Tel: 0964 533878

**Nobby Clark British
Motorcycles**
Unit 2, The Old Tannery,
The Midlands, Holt,
Wiltshire, BA14 6RW, UK

Tel: 0225 782923

Norman Hyde
Rigby Close, Heathcote
Ind.Estate, Heathcote,
Warwickshire, CV34 6TL,
UK

Tel: 0926 497375

**Norman Watt
Motorcycles**
151 Saintfield Road, Temple
Cross Roads, Boardmills,
Lisburn, Co.Down, PT26
6UG, UK

Tel: 0846 638766

**North Harbour
Motorcycles**
14-18 North Harbour
Street, Ayr, Strathclyde,
KA8 8AA, UK

*Contact: Mr Jim Neal
Tel: 0292-281933
Fax: 0292-281933*

Officially appointed dealers for
BMW, Honda, Suzuki, Yamaha
and Kawasaki. Large range of
new and used motorcycles in
stock. Clothing department,
parts, service and insurance.

Northsport Motorcycles
48-54 Bishopton Lane,
Stockton on Tees,
Cleveland, TS18 2AQ, UK

Tel: 0642 604810

Norton Motors
Lynn Lane, Shenstone,
Lichfield, Staffordshire,
WS14 0EA, UK

Tel: 0543 480101

Orwell Cycles
Clarkson Street, Ipswich,
Suffolk, IP1 2NB, UK

Tel: 0473 257401

**Oxford M/C Engineers
Ltd**
10 Hythe Bridge Street,
Oxford, Oxfordshire,
OX1 2EW, UK

Tel: 0865 250570

P & H Motorcycles
6-7 Orchard Street,
Crawley, Sussex (W.), UK

Tel: 0293 519465

P & R Motorcycles
35/37 Bradford Road,
Stanningley, Pudsey,
Yorkshire (W.), UK

Tel: 0532 552346

P.R. Taylor & Sons
23-25 Station Hill,
Chippenham, Wiltshire,
SN15 1EG, UK

Tel: 0249 657575

PGH Motorcycles
160-162 Avenue Road,
Torquay, Devonshire,
TQ2 5LQ, UK

Tel: 0803 212161

PW Ranger
40/41 Gaol Road, Stafford,
Staffordshire, ST16 3AR,
UK

Tel: 0785 47190

Parks Of Lewisham
404 Lewisham High Street,
Lewisham, London,
SE13 6LS, UK

Tel: 081 690 8666

Paul Hunt Motorcycles
Alfreton, Derbyshire, UK

Tel: 0773 540526

Pegasus Kawasaki Centre
324 Oxford Road, Reading,
Berkshire, RG3 1AY, UK

Tel: 0734 571977

Phil Cotton Classic Bikes
2 King Street,
Dalton-in-Furness,
Cumbria, LA15 8JA, UK

Tel: 0229 467457

Powerslide Bikes
1 High Street, Woodville,
Swadlincote, Derbyshire,
DE11 7EH, UK

Tel: 283 551001

QB Motorcycles
89-91 High St., Quarry
Bank, nr.Merry Hill Centre,
Midlands (W.), 075 2AD,
UK

Tel: 0384 637168
Fax: 0384 411021

Honda, Yamaha, Cagiva and
Moto Guzzi main dealer's
service, repairs and MOTs. Full
clothing and accessories
department. Finance arranged,
part exchange welcome. Open
Mon-Sat, 11-2pm Sun.

Queens Park Motors Ltd
13 & 21 Ford Lane, Salford,
Manchester, Gt., M6 6PE,
UK

Tel: 061 736 2585

R A Wilson Motorcycles
Unit 32, London Rd.
Ind.Estate, London Road,
Grantham, Lincolnshire,
NG31 6HP, UK

Tel: 0476 593793

R Judd Ltd
415 Burnt Oak Broadway,
Edgware, London, HA8
5AH, UK

Tel: 081 952 6911

R W Parkinson & Son
35-37 London Road, Marks
Tey, Colchester, Essex,
CO6 1DZ, UK

Tel: 0206 210467

**RAP International
Superbikes**
2 Cricklewood Broadway,
London, NW2 2BT, UK

Tel: 081 452 2672

Rafferty Newman Ltd
32 Exmouth Road,
Southsea, Portsmouth,
Hampshire, PO5 2QL, UK

Tel: 0705 755125

Open daily 9.15am to 6.00pm for
Yamaha, Kawasaki, Triumph,
Suzuki, Honda, Vespa, Mig and
Lambretta. Over the counter or
mail order, spares machines or
accessories.

Rainbow Motorcycles
160 Broad Oaks, Sheffield,
Yorkshire (S.), S9 3HJ, UK

Tel: 0742 618866

Reg Allen
37/41 Grosvenor Rd,
Hanwell, London, W7, UK

Tel: 081 567 1974
Fax: 081 579 1248

London's largest stockist of
Meriden Triumph spares;
machines bought, sold and
exchanged.

Regent House
Neatgangs Lane, Goxhill,
Humberside (S.),
DN19 7NL, UK

Tel: 0469 32251

Revettes Ltd
53-67 Norwich Road,
Ipswich, Suffolk, IP1 2ER,
UK

Tel: 0473 253726

Riders (Bridgewater) Ltd
Riders House, Wylds Road,
Bridgewater, Somerset,
TA6 4BH, UK

Tel: 0278 457652

Road Star Cycles
Victoria Park mews, Dover,
Kent, CT16 1QS, UK

Tel: 0304 202881

**Rob Willsher
Motorcycles**
Lowford Hill, Bursledon,
Southampton, Hampshire,
SO3 8ES, UK

Tel: 0703 403203
Fax: 0703 405087

Suzuki specialists for over 20
years.

Robert Bevan & Son
29/33 Castle Street, Cardiff,
Glamorgan (S.), CF1 2BT,
UK

Tel: 0222 227477

**Robertsbridge Classic
MCs**
Western House, Station
Road, Robertsbridge,
Sussex (E.), TN32 5DE, UK

Tel: 0580 880323

Robinsons
The Foundry, Broad Oak
Road, Canterbury, Kent,
CT2 7QG, UK

Tel: 0227 45366
Fax: 0227 454726

Harley-Davidson and Suzuki
main agents. Monday to Friday
9am to 5.30pm, Sat 9am to 5pm.
MOT, mail order spares, clothing
dept.

Robinsons of Rochdale
Central Garage, Water
Street, Rochdale,
Manchester, Gt.,
OL16 1UH, UK

Tel: 0706 45964

Roy Pidcock Motorcycles
277 Osmaston Road,
Derby, Derbyshire, UK

Tel: 0332 49673

Rye's of Southampton
West Quay Road,
Southampton, Hampshire,
SO1 0GZ, UK

Tel: 0703 321321

Honda main dealer open
8.00am-5.30pm, Mon-Sat. Full
spares and clothing display.
MOT by appointment.

SEP
39 Sidelay Rd, Kegworth,
Derbyshire, DE7 2FJ, UK

Tel: 0509 673295

SGT Superbiking
351 Bath Road, Slough,
Berkshire, SL1 5PR, UK

Tel: 0753 811122

**Saddleworth Classic
Motorcycles**
Knarr Mill, Oldham Road,
Delph, Lancashire,
OL3 5RQ, UK

Tel: 0457 872788

Saunders Motorcycles
17 Station Road,
Knebworth, Hertfordshire,
SG3 6AP, UK

Tel: 0438 811524

Scotbike Ltd
250 Great Western Road,
Glasgow, Strathclyde,
G4 9EJ, UK

Tel: 041 331 1199
Fax: 041 332 8474

Closed Sun and Mon, open late
Thurs. New motorcycle
warehouse now open in
Glasgow. Phone for details.

**Scotland Alvins
Motorcycles**
9A-9B Springfield Street,
Edinburgh, Lothian,
EH6 5EF, UK

Tel: 031 555 1039

Sherwood Garage
194 Robin Hood Lane, Hall
Green, Birmingham,
Midlands (W.), B28 0LG,
UK

Tel: 021 777 1311

Shirlaws Motorcycles
92 Crown Street,
Aberdeen, Grampian,
AB1 2HJ, UK

Contact: Roy Shirlaw
Tel: 0224 584855

Mail order on bikes by BMW, Ducati, Triumph, Yamaha, Suzuki, Honda, Vespa. Parts, clothing & helmets, specialist repairs, training. Motorcycle world all under one roof.

Simpson Mecanique
2 Bis Rue de Corneilhan,
Beziers, 34500, France

Simpson Mecanique Ducati
Routes de Saint Georges,
Juvignac, 34990, France

Tel: 010 33 670 30700
Fax: 010 33 670 30945

Modern Ducati dealer, has dynamic wide-case (post '68) single spares department with massive stocks of racing parts and accessories. English catalogue available. Stocks parts for all V-Twins too.

Skellerns Motorcycles
52 Sidbury, Worcester,
Hereford & Worcester,
WR1 2NA, UK

Tel: 0905 20580

Triumph, Honda, Kawasaki and Yamaha motorcycles, parts, helmets, clothing, servicing and MOTs. Open Tuesday-Saturday. Sales: 0905-20580. Parts/Clothing: 0905-23254. Service: 0905-724848

Slinger Motorcycles
Waterloo Garage,
40 Waterloo Road,
Preston, Lancashire,
PR2 1BQ, UK

Tel: 0772 727213

Solihull Motorcycles
59 Hobs Moat Road,
Solihull, Midlands (W.), UK

Tel: 021 742 1414

South Essex Motorcycles
15-17 Southend Road,
Grays, Essex, RM17 5NH,
UK

Tel: 0375 375653

South Wales Superbikes
Redcastle Garage, Malpas
Road, Newport, Gwent,
NP1 6WB, UK

Tel: 0633 821100

Southern Motorcycles
47-65 Bevois Valley Road,
Southampton, Hampshire,
SO2 0JS, UK

Contact: Peter Edwards
Tel: 0703 235834
Fax: 0703 235834

Used motorcycles of quality. Full spares and repair facilities open seven days a weeek. Fixed price servicing, full range of American imports.

Southsport Superbikes
86-90 Eastbank Street,
Southport, Merseyside,
PR8 1EF, UK

Tel: 0704 536192

Speedway Motorcycles
2 Belgrave Road,
Halesowen, Midlands (W.),
UK

Tel: 021 559 1270

Sportbike
81-83 Stoke Road, Shelton,
Stoke on Trent,
Staffordshire, ST4 2QH, UK

Tel: 0782 415768

St Lukes Motorcycles
71 London Street,
Southport, Merseyside,
PR9 0TX, UK

Tel: 0704 542218

St Neots Motorcycle Company
29-39 St Marys Road,
Eynesbury, St Neots,
Cambridgeshire, PE19 2TA,
UK

Tel: 0480 212024

St Peter Port Garages Ltd
Trinity Square, St Peter
Port, Guernsey, CI, UK

Tel: 0481 724261

Staccato Ducati
57 Primrose Corner,
Woodbastwick, Norwich,
Norfolk, NR13 6JL, UK

Tel: 0603 721493

Stevens & Stevens Trials Centre
Unit 43, Blue Chalet Ind.
Park, London Road,
West Kingsdown, Kent,
TN15 6BQ, UK

Tel: 0474 854265
Fax: 0474 854032

Trials motorcycles, spares, accessories & clothing. Advice on joining clubs (adults & juniors).

Stibbs
25 Golden Hill,
Wiveliscombe, Somerset,
TA4 2NT, UK

Tel: 0984 23956

Street Bike
73 King Street, Dudley,
Midlands (W.), DY2 8QB,
UK

Tel: 0384 253464

Street Bike
3 The Ringway, West
Bromwich, Midlands (W.),
B70 6EU, UK

Tel: 021 525 1429

Streetbike (Dudley)
73 King Street, Dudley,
Midlands (W.), DY2 8QE,
UK

Tel: 0384 253464

Sturdey Motorcycles
94a-94c Shipbourne Road,
Tonbridge, Kent,
TN10 3EG, UK

Tel: 0732 354082

Sumo Bikes
Arch 6, Adenmore Road,
Catford, London, SE6, UK

Tel: 081 314 0468

Surrey Harley-Davidson
Unit 5, Havenbury Estate,
Station Road, Dorking,
Surrey, RH4 1ES, UK

Tel: 0306 883825

Suzuki/Information Dept.
PO Box 56, Tunbridge
Wells, Kent, TN1 2XY,
UK

Tel: 0500 526262

TW Motorcycles
139 Oxford Road,
Kidlington, Oxfordshire,
OX5 2NP, UK

Tel: 086 75 5702

TT Motorcycles
Grigor Hill Ind Estate,
Nairn, Highland, IV12 5HY,
Scotland

Tel: 0667 53540
Fax: 0667 52837

Taylor Racing
23-25 Station Hill,
Chippenham, Wiltshire, UK

Tel: 0249 657575/6

The Bike Connection
58-60 Dallow Road, Luton,
Bedfordshire, LU1 1LY, UK

Tel: 0582 423614

The Foundry
Broad Oak Road,
Canterbury, Kent,
CT2 7QG, UK

Tel: 0227 463986

The Harley Shop
Harley House, Whitwood
Common Lane, Castleford,
Yorkshire (W.),
WF10 5PD, UK

Tel: 0977 517566

**Three Cross
Motorcycles Ltd**
Woolsbridge Ind.Estate,
6 Old Barn Farm Road,
Three Legged Cross,
Wimbourne, Dorset,
BH21 6SP, UK

Tel: 0202 824531

Thruxton Motorcycles
Station Approach, Andover,
Hampshire, SP10 3HN, UK

Tel: 0264 354200

'Hampshire's sporting dealer' -
agents for Honda, Kawasaki,
Suzuki, Yamaha, Ducati; plus
secondhand machines, servicing,
repairs, part-exchange, finance,
MOTs. Our own all-activity
motorcycle club.

Tillstons Ltd
52 Yarn Lane, Stockton on
Tees, Cleveland, TS18 1ER,
UK

Tel: 0642 611138

Tinklers Motorcycles
182-190 Northumberland
Street, Norwich, Norfolk,
NR2 4EE, UK

Tel: 0603 627786

Tippetts Motors Ltd
312-320 Ewell Road,
Surbiton, Surrey,
KT6 7AW, UK

Tel: 081 399 2417
Fax: 081 390 6759

Honda motorcycle sales, spares
and service. Open
8.30am-5.45pm Monday-Friday,
8.30am-4.45pm Saturday. Credit
cards and telephone orders
accepted.

**Tommy Robb
(Motorcycles) Ltd**
240 Manchester Road,
Warrington, Cheshire,
WA1 3BE, UK

Tel: 0925 56528
Fax: 0925 234518

Main agents for Honda,
Triumph, Kawasaki, Yamaha,
Piaggio motorcycles. We offer a
vast range of accessories:
clothing, helmets, boots, gloves,
etc.; plus advice on training,
insurance, finance etc. Our
service department is the best for
repairs and MOTs.

Tony Grant
25 Orchard Road,
Stevenage, Hertfordshire,
SG1 3HE, UK

Tel: 0438 315562

Townsends Motorcycles
57 Stafford Road,
Wallington, Surrey, UK

Tel: 081 647 5393

Tri-Supply
2 Kemp Road, Winton,
Bournemouth, Dorset,
BH9 2PW, UK

Tel: 0202 514446

Triumph In London
151 Plumstead Road,
London, SE18 7DY, UK

Tel: 081 854 8133

**Triumph Motorcycles
Ltd**
Jacknell Road, Dodwells
Bridge Ind. Estate, Hinckley,
Leicestershire, LE10 3BS,
UK

Tel: 0455 251700

Twiggers Motorcycles
30 Nottingham Road,
Loughborough,
Leicestershire, LE11 1EU,
UK

Tel: 0509 263967

Honda motorcycles stocked in
Leicestershire (off Junction
23-M1). Twiggers Motorcycles
for new Honda sales, good used
motorcycles bought and sold.
Service work in Honda
approved workshop, repairs,
insurance and spares for most
makes. MOTs, tyres supplied,
fitted, balanced and repaired.

Two Wheel Services Ltd
79 Nolton Street, Bridgend,
Glamorgan, Mid-,
CF31 3AE, UK

Tel: 0656 657887

Two Wheel Tyre Centre
6 Queslett Road, Great
Barr, Birmingham, Midlands
(W.), B43 6PL, UK

Tel: 021 357 3998

Two Wheels
35 Cross Street,
Farnborough, Hampshire,
UK

Tel: 0252 522552

Verralls (Handcross) Ltd
The Old Garage, High
Street, Handcross, Sussex
(W.), RH17 6BJ, UK

Contact: *Brian R Verrall*
Tel: 0444 400 678
Fax: 0444 401 111

Dealers in vintage, veteran and
classic motorcycles open
Tues-Sat, 10am-5pm. Catalogue
available (please send suitable
postage coupons).

**Vin Duckett Motorcycle
Centre**
Anchorsholme Lane East,
Blackpool, Lancashire,
FY5 3QJ, UK

Tel: 0253 826142

W E Wassell
4 Little Barr Street,
Birmingham, Midlands (W.),
B9 4HE, UK

Tel: 021 773 2837

WeeVee
237 London Road,
Croydon, Surrey, UK

Tel: 081 684 2869

**Wheelpower Bike
Centre**
264 Grand Drive, Raynes
Park, London, SW20 9NE,
UK

Tel: 081 543 0321

**Wheelpower Bike
Centre**
849 Fulham Road, London,
SW6, UK

Tel: 071 736 7965

Wheels International
Watling Street (A5),
Hockliffe, Bedfordshire,
LU7 9LS, UK

Tel: 0525 210130

Wilson
P.O.Box 88, Crewe,
Cheshire, CW1 YZ, UK

Tel: 0270 668523

Windy Corner
8 Moat Way, Barwell,
Leicestershire, LE9 8EY, UK

Tel: 0455 842922

Woods Motorcycles
Rhuddian Road, Abergele,
Clwyd, LL22 7HF, UK

Tel: 0745 825958

Wylie & Holland
63-67 Shrewsbury Road,
Market Drayton,
Shropshire, TF9 3DN, UK

Tel: 0630 657121

York Suzuki Centre
179-181 Burton Stone Lane,
Clifford, Yorkshire (W.),
YO3 6DG, UK

Tel: 0904 625404

INSURANCE

Apex Insurance
10 Park Street, Selby,
Yorkshire (N.), YO8 0PW,
UK

Tel: 0757 709121

Bennetts Insurance Brokers
12 Ironmonger Row,
Coventry, Midlands (W.),
CV1 1FD, UK

Tel: 0203 553221
Fax: 0203 520790

Wide range of competitive motorcycle insurance from the country's largest independent motorcycle insurance brokers - schemes available for female riders - despatch riders - Suzuki owners. Group rates or rider policies and six months policies for "Summer only" bikers.

British Bike Insurance
KGM House, George Lane,
London, E18 1RF, UK

Tel: 081 530 6252

Carole Nash Insurance Consultants Ltd
Paul House,, Stockport
Road, Timperley,
Altrincham, Cheshire,
WA15 7UQ, UK

Tel: 061 980 1305
Fax: 061 903 9784

Carole Nash insurance consultants are the country's leading specialist motorcycle insurers for both modern and classic motorcycles, with a number of different schemes available.

Compensation Direct
Unit 4, Hollbrook Arcade,
Southampton General
Hospital, Southampton,
Hampshire, UK

Tel: 0703 510011

D & J Davies (Brokers) Ltd
11 North Bridge Street,
Shefford, Bedfordshire, UK

David Levene & Co
560-568 High Road,,
London, N17 9TA, UK

Tel: 081 365 1822

Devitt Ltd
Central House, 32-66 High
Street, Stratford, London,
E15 2PF, UK

Tel: 081 555 0711

Footman James & Co Ltd
Waterfall Lane, Cradley
Heath, Warley, Midlands
(W.), B64 6PU, UK

Tel: 021 561 6222
Fax: 021 561 4093

Classic/modern motorcycle insurance with limited/unlimited mileage options - write, phone or fax for a quotation.

ITT Commercial Finance Ltd
Victoria Gate, Chobham
Road, Woking, Surrey,
GU21 1JU, UK

Tel: 0483 727788

Kestral Insurance Services
12 Dock Street, Hull,
Humberside (N.),
HU1 3DL, UK

Tel: 0482 210012

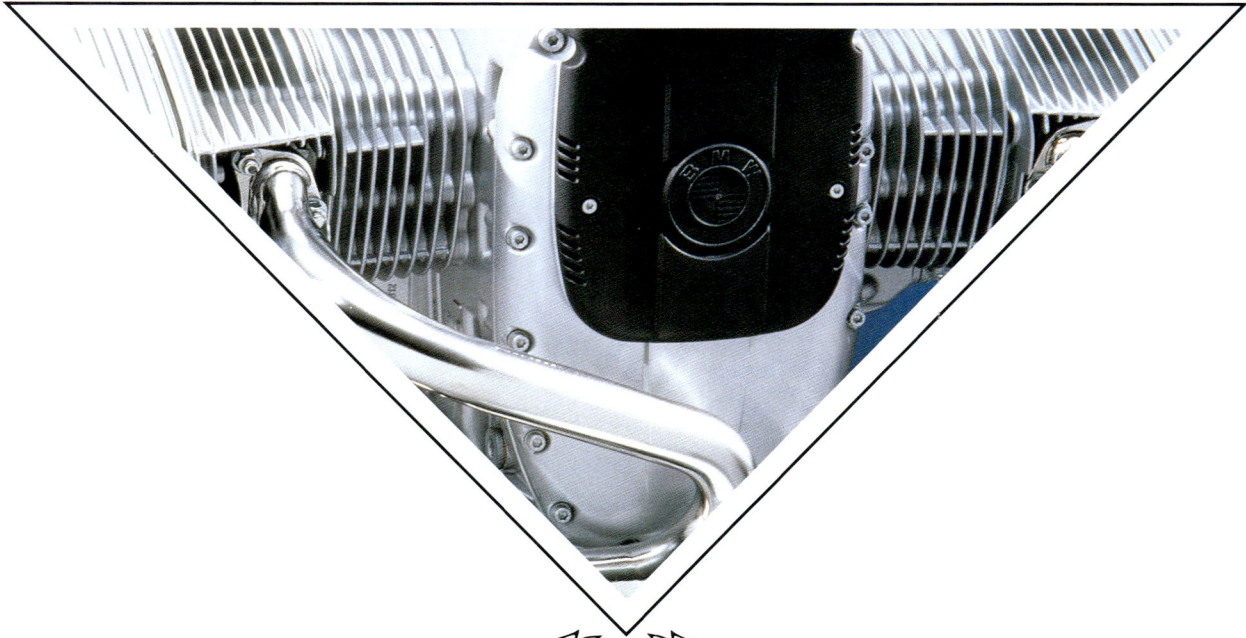

Kirwans Solicitors
363 Woodchurch Road,
Birkenhead, Merseyside,
L42 8PE, UK

Contact: *Claims Manager*
Tel: 051 608 9078

Mitchell & Partners
15 Archway Road, London,
N19 3TV, UK

Tel: 071 281 7661

NLVR
P.O.Box 1455, Windsor,
Berkshire, SL4 1QR, UK

Tel: 0753 831553

Norwich Union Fire Insurance
PO Box 4, Surrey Street,
Norwich, Norfolk,
NR1 3NG, UK

Tel: 0603 622200

Osborne & Sons
2 Rose Hill, Sutton, Surrey,
SM1 3EU, UK

Tel: 081 641 6633
Fax: 081 641 1894

RIGP Finance Ltd
62 Player Street, Radford
Boulevard, Nottingham,
Nottinghamshire,
NG7 2FY, UK

Tel: 0602 420222

Slocombes
251-3 Neasden Lane,
London, NW10 1QH, UK

Tel: 081 450 6644

Swann & Moore (Assessors) Ltd
83a St John's Way,
Corringham,
Stanford-le-Hope, Essex,
SS17 7LL, UK

Tel: 0375 640166
Fax: 0375 644098

Accident Claims Consultants
operating on "No Win, No Fee"
basis. 30 years experience
handling claims for
compensation arising from road
accidents. Motorcyclists a
speciality.

Transamerica Commercial Finance Co.Ltd
18 Market Place, Hitchin,
Hertfordshire, SG5 1DS,
UK

Tel: 0462 420442

Whittlesey Insurance Services Ltd
12 Queens Street,
Whittlesey, Peterborough,
Cambridgeshire, PE7 1AY,
UK

Tel: 0733 208117

Registered insurance brokers
transacting motorcycle insurance
and all other forms of insurance:
including private car, buildings
and contents, commercial covers
and life and pension advice.

MAGAZINES & PUBLISHERS

ACU Newsline
Wood Street, Rugby,
Warwickshire, CV21 27X,
UK

Tel: 0788 540519

AWOL
P.O.Box 122, Northwich,
Cheshire, CW9 8YA, UK

Back Street Heroes
P.O.Box 28, Altrincham,
Cheshire, Wa15 8SH, UK

Bike Magazine
20-22 Station Road,
Kettering,
Northamptonshire,
NN15 7HH, UK

Contact: Martyn Moore
Tel: 0536 416416
Fax: 0536 415748

Britain's biggest monthly
motorcycle magazine. All that's
fast, furious and fun on two
wheels. On sale on the 8th of
every month.

British Bike Magazine
Green Designs, PO Box 19,
Cowbridge, Glamorgan (S.),
CF7 CYD, UK

Tel: 0446 775033
Fax: 0446 772204

The interesting, practical,
inspiring magazine for classic
bike owners - incorporates Ace
magazine, a regular supplement
which deals with Italian,
American and Japanese bikes.

Classic Bike Guide
Myatt McFarlane,
PO Box 666, Altrincham,
Cheshire, WA15 2UD, UK

Tel: 061 928 3480

Classic Mechanics
Suite G, Deene House,
Market Square, Corby,
Northamptonshire,
NN17 1PB, UK

Classic Motorcycle
Bushfield House, Orton
Centre, Peterborough,
Cambridgeshire, PE2 5UW,
UK

Contact: **June Luxford**

**Classic Motorcycling
Legends**
D.C.Thomson & Co,
Dundee, Tayside,
DD1 9QJ, UK

Tel: 0382 23131

**Condor Motorcycle
Developments**
Warwick House, 59/60
Derby Square, Douglas,
Isle Of Man, UK

Dirt Bike Rider
Key Publishing Ltd,
PO Box 100, Stamford,
Lincolnshire, PE9 1XQ, UK

Fast Bikes
Studio 12, 92 Lots Road,
Chelsea, London,
SW10 OQD, UK

Going Places
BBC Radio 4, Broadcasting
House, London,
W1A 1AA, UK

Heavy Duty
Power Play Publications,
Parker House, Tan Yard
Lane, Bexley, Kent,
DA2 1AH, UK

Tel: 0322 555959

J H Haynes & Co Ltd
Sparkford, Nr.Yeovil,
Somerset, BA22 7JJ, UK

Tel: 0963 40635

Haynes produce over 150 titles
for the motorcyclist, ranging
from the ever popular Owners
Workshop Manuals series to
many general motorcycling
books.

**JR Technical
Publications Ltd**
Potterdike House, Lombard
Street, Newark,
Nottinghamshire,
NG24 1XG, UK

London Biker
Publications In Business Ltd,
30 Rathbone Place, London,
W1P 1AD, UK

Tel: 071 323 6926

Magnews
MAG Central Office,
P.O.Box 750, Birmingham,
Midlands (W.), B30 3BA,
UK

Metal And Leather
PO Box 28, Altrincham,
Cheshire, WA15 8SH, UK

Tel: 061 929 1332

Motocross Rider
Sandylands House,
Sandylands, Morecambe,
Lancashire, LA3 1DG, UK

Motorcycle Check Book
Glass's Guide Service Ltd,
Elgin House, St.George's
Ave., Weybridge, Surrey,
KT13 0BX, UK

Tel: 0932 853211

Motorcycle Dealer
P.O.Box 11, 20/22 Station
Road, Kettering,
Northamptonshire,
NN15 7TG, UK

Motorcycle International
Myatt McFarlane PLC,
PO Box 666, Altrincham,
Cheshire, WA15 2UD, UK

Tel: 061 928 3480

Motorcycle News
Abbots Court,
34 Farringdon Lane,
London, EC1R 3AU

Tel: 071 837 3699

Motorcycle Rider
P.O.Box 17, Hertford,
Hertfordshire, SG14 1SG,
UK

Motorcycle Rider (BMF)
129 Seaforth Avenue,
Motspur Park, New
Malden, Surrey, KT3 6JU,
UK

Tel: 081 942 7914

Motorcycle Sport
Ravenhill Publishing,
Standard House, Bonhill
Street, London,
EC24 4DA, UK

Tel: 071 628 4741

Motorcycle Trader
Seven Kings Publications
Ltd, Garden Hall Ho.,
Wellesley Rd., Sutton,
Surrey, SM2 5UF, UK

Tel: 081 661 1160

Norton Owners Club
30 Rosehill Avenue, Sutton,
Surrey, SM1 3HG, UK

Old Bike Mart
P.O.Box 6,
Chapel-en-le-Frith,
Stockport, Cheshire,
SK12 6LD, UK

Performance Bikes
EMAP Ltd, Bushfield House,
Orton Centre,
Peterborough,
Cambridgeshire, PE2 5UW,
UK

Tel: 0733 237111

RPM
Pinegen Ltd, 10A Station
Road, Cuttley, Hertford-
shire, EN6 4HT, UK
Tel: 0707 876219

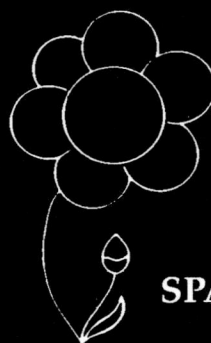

Road Racing Monthly
Huthwaite Printing Co,
Common Road, Sutton in
Ashfield, Nottinghamshire,
NG17 6AA, UK

Scooter Scene Magazine
P.O.Box 46, Weston super
Mare, Avon, BS23 1AF, UK

Scootering
P.O.Box 69, Altrincham,
Cheshire, WA15 5SJ, UK

Silver Machine
Advanced Publishing, 32
Paul Street, London, EC2A
4LB, UK

Streetfighters
Myatt McFarlane, PO Box
666, Altrincham, Cheshire,
WA15 2UD, UK

Tel: 061 928 3480

SuperBike
Link House, Dingwall
Avenue, Croydon, Surrey,
CR9 2TA, UK

TRF Bulletin
6 Glasgow Road, Biterne,
Southampton, Hampshire,
SO2 5PE, UK

Top Gear
BBC Pebble Mill,
Birmingham, Midlands (W.),
BT 7QQ, UK

**Trials and Motocross
News**
112 Victoria Street,
Morecambe, Lancashire,
LA4 1DG, UK

Used Bike Guide
Bradley Publishing Services,
Parker House, Tan Yard
Lane, Bexley, Kent,
DA5 1AH, UK

What Bike
Bushfield House, Orton
Centre, Peterborough,
Cambridgeshire, PE2 5UW,
UK

MANUFACTURERS

AJS Motorcycles
Goodworth, Clatford,
Andover, Hampshire,
SP11 7RP, UK

Tel: 0264 710548

Acrybre Products
10 Albany Road, Granby
Industrial Estate,
Weymouth, Dorset,
DT4 9TH, UK

Contact: *Mr B Wightman*
Tel: 0305 787498
Fax: 0305-787499

Manufacturers of motor cycle
windscreens and fibre glass
accessories.

Allwick Patterns
Seager Road, Oare,
Faversham, Kent,
ME13 7TL, UK

Tel: 0795 532580

Anglo German Tools
Unit C3, Phoneix Tdg.
Estate, Thrupp, Stroud,
Gloucestershire, GL5 2BQ,
UK

Tel: 0453 886258

Aprilia Moto UK Ltd
Unit 11, Gregory Way,
South Reddish, Stockport,
Cheshire, SK5 7ST, UK

Tel: 061 476 5770

Ariel Dragonfly
The Old Town Maltings,
Broad Street, Bungay,
Suffolk, NR35 1EE, UK

Tel: 0986 894798

Autovalues Engineering
Albion Road, Idle,
Bradford, Yorkshire (W.),
BD10 9RL, UK

Tel: 0274 614424

Ava Europa Ltd
Dunragit, Wigtown,
Dumfries & Galloway, DG9
8PN, UK

Tel: 05814 636

Avon Tyres Ltd
Bath Road, Melksham,
Wiltshire, SN12 8AA, UK

Tel: 0225 703101

B & C Express Products
Station Road,
Potterhanworth,
Lincolnshire, LN4 2DX, UK

Tel: 0522 791369

B G Hichisson
Vauxhall Place, Lowfield
Street, Dartford, Kent,
DA1 1HU, UK

Tel: 0322 220674

Light alloy petrol tanks for
vintage motor cycles, BSA,
Featherbed Nortons, Ariel &
AMC competition models.

BLR Engineering
1 Paragon Grove,
Berrylands Road, Surbiton,
Surrey, UK

Tel: 081 399 6617

BMW (GB) Ltd
Ellesfield Avenue, Bracknell,
Berkshire, RG12 4TA, UK

Tel: 0344 426565

Baby Biker
95 Gretna Green, Green
Lane, Coventry, Midlands
(W.), CV3 6DT, UK

Tel: 0203 417532

Baglux UK
75 Foryd Road, Kinmel Bay,
Rhyl, Clwyd, LL18 5BB, UK

Tel: 0745 338080
Fax: 0745 337339

Baglux tailor made protective
colour matched tank covers and
tank bags designed to fit your
machine perfectly. Available to
fit nearly all machines from 1979
to 1994. New custom leather
panniers, saddle bags, seat and
tool bags, nylon seat bags,
handle bar muffs and machine
aprons. Available from over 30C
UK dealers. Phone for further
details.

Basset Down Balancing
Unit 19, Lower Bassett
Down, Swindon, Wiltshire,
SN4 9QR, UK

Tel: 0793 812331

Bavanar Products Ltd
47 Beaumont Road, Purley,
Surrey, CR2 2EJ, UK

Tel: 081 645 9190

Bek Wholesale Company
Unit 1, Light Industrial Units,
Station Road, Worthing,
Sussex (W.), BN11 1LP, UK

Tel: 0903 821376

Bert Harkins Racing
Unit 6, Townsend Centre,
Houghton Regis, Dunstable,
Bedfordshire, UK

Tel: 0582 472374

Bifax International Ltd
Royce Close, Andover,
Hampshire, SP10 3TS, UK

Tel: 0264 361411

**Bike Tech (BMS
Developments)**
104 Richmond Park Road,
Bournemouth, Dorset,
BH8 8TH, UK

Tel: 0202 33370

Bill Head (Preston) Ltd
Southgate, Preston,
Lancashire, PR1 1NP, UK

Tel: 0772 52066

Bob Heath Visors Ltd
6 Birmingham Road,
Walsall, Midlands (W.),
WS1 2NA, UK

Tel: 0922 614747

Boretech Engineering
Unit 10, Golding Barn Race-
way, Henfield Road, Small
Dole, Sussex (W.), UK

Tel: 0903 816236

Brian Bennett
13 Derby Road Business
Park, Derby Road, Burton
on Trent, Staffordshire,
DE14 1RW, UK

Tel: 0283 511841

**Bridgestone/Firestone
UK Ltd**
Bridgestone House,
Birchley Trading Estate,
Oldbury, Warley, Midlands
(W.), B69 1DT, UK

Tel: 021 552 3331

**British Motorcyclists
Federation**
129 Seaforth Avenue,
Motspur Park, New
Malden, Surrey, KT3 6JU,
UK

Tel: 081 942 7914

C I Sport
205-210 Kingston Road,
Leatherhead, Surrey,
KT22 7PB, UK

Tel: 0372 378000

CHL Engineering Ltd
86 Herbert Road,
Plumstead, London,
SE18 3TD, UK

Tel: 081 855 1234

CPK Auto Products Ltd
Gladiator Way, Glebe Farm
Estate, Rugby,
Warwickshire, CV21 1PX,
UK

Tel: 0788 540033

Castrol UK Ltd
Burmagh Castrole House,
Pipers Way, Swindon,
Wiltshire, SN3 1RE, UK

Tel: 0793 512712

Chassis Dynamics
(See our display
advertisement, or call us on
number below),

Tel: 0604 22252

**Continental Tyre &
Rubber Group Ltd**
4-8 High Street, West
Drayton, Middlesex,
UB7 7OJ, UK

Tel: 0895 445678
Fax: 0895 431508

CTRG is the UK sales and
marketing company for the
Continental brand of motorcycle
tyres, including a range of radial,
bias belted and crossply tyres for
numerous applications.

Cougar Customs
Upper Polmaise, Stirling,
Grampian, FK7 9PU, UK

Tel: 0786 465778

Motoliner frame & fork
straightening, custom frames,
tanks, yoke & wide glide kits,
exhaust fabrications, milling,
turning, alloy welding. Open
Mon-Sat, 9-5.

Crafty Toys
8a Douglas Road,
Maidstone, Kent,
ME16 8ES, UK

Tel: 0622 755293

Crusader Leathers Ltd
544 Sheffield Road,
Whittington Moor,
Chesterfield, Derbyshire,
S41 8LX, UK

Tel: 0246 454921

D J Allwood
R/O Arch 354, Winchelsea
Road, Forest Gate,
London, E7 0AG, UK

Tel: 081 555 9118

David Silver Spares
Unit 14, Masterlord Ind.Est.,
Station Road, Leiston,
Suffolk, IP16 4JD, UK

Tel: 0728 833020

Dawson Harmsworth Ltd
401-403 Penistone Road,
Sheffield, Yorkshire (S.),
S6 2FL, UK

Tel: 0742 337460

Delta Clothing
46A Gladstone Street,
Bedford, Middlesex,
MK41 7RR, UK

Tel: 0234 52221

**Dennis Trollope Racing
Spares**
131 Station Road,
Kingswood, Bristol, Avon,
BS15 4XX, UK

Tel: 0272 570821

Devon Rim Company
20 South Street, South
Molton, Devonshire,
EX36 4AG, UK

Contact: *Doug Richardson*
Tel: 0769 574108
Fax: 0769 574436

Manufacturers of high quality
stainless steel wheel rims which
look like chrome but last forever.
Stainless spokes and Avon tyres
supplied. Also full
wheelbuilding service.

Dynamic Ltd
Dynamic House, Spring
Road, Walsall, Midlands
(W.), WS4 1QQ, UK

Tel: 0922 693946

Feridax (1957) Ltd
Park Lane, Halesowen,
Midlands (W.), B63 2NT,
UK

Tel: 0384 410384

Ferodo Ltd
Chapel-en-le-Frith,
Stockport, Cheshire,
SK12 6JP, UK

Tel: 0298 812520

Fieldsheer Europe
10 Penchos Avenue, Gatley,
Stockport, Cheshire, UK

Tel: 061 428 5401

**Freeman Automotive
UK Ltd**
Units B-H Barker Buildings,
Countess Road, Spencer
Bridge, Northamptonshire,
NN5 7EA, UK

Tel: 0604 583344

Freewheel UK Ltd
Cornish Way, North
Walsham, Norfolk,
NR28 0AW, UK

Contact: *Alan Young*
Tel: 0692 500300

Freewheel manufacture:
luggage, trailers, turnbars, bike
trailers, top brakes, replacement
clutches, panniers, racks and
custom made parts for
motorcycles; from steel, stainless
and G.R.P.

G K Blair
Cliff Cottage, Marchington
Woodlands, Uttoxeter,
Staffordshire, ST14 8PB,
UK

Tel: 0283 820508

HT Engineering
350 Leatherhead Road,
Chessington, Surrey,
KT9 2NN, UK

Tel: 0372 740306

Harley-Davidson UK Ltd
The Bell Tower, High
Street, Brackley,
Northamptonshire,
NN13 5DT, UK

Tel: 0280 700101

Harman Services
Unit 1, Little Park
Enterprise, Ifield, Crawley,
Sussex, RH11 0BW, UK

Tel: 0293 513266

**Harris Performance
Prods.Ltd**
6 Marshgate Ind Est,
Hertford, Hertfordshire,
SG13 7AQ, UK

Tel: 0992 551026

Harrison Engineering
Unit 35, Blue Chalet
Ind.Park, London Road,
West Kingsdown, Kent,
TN15 6BT, UK

Hein Gericke (UK) Ltd
35 Blossom Street, York,
Yorkshire (N.), YO2 2AQ,
UK

Tel: 0904 655379

Heron Suzuki Plc
46/62 Gatwick Road,
Crawley, Sussex (W.),
RH10 2XF, UK

Tel: 0293 518000

Hill Marketing Ltd
3 Omni Business Centre,
Omega Park, Alton,
Hampshire, GU34 2QD,
UK

Tel: 0420 541444

Honda Motor Europe Ltd
4 Power Road, Chiswick,
London, W4 5YT, UK

Tel: 081 746 9231

J & S Accessories
Unit 9, Denton Drive,
Northwich, Cheshire,
CW9 7LU, UK

Tel: 0606 48691

**Kawasaki Motors (UK)
Ltd**
1 Dukes Meadow,
Millboard Road, Bourne
End, Buckinghamshire,
SL8 5XF, UK

Tel: 0628 851000

Lintex Gleave Ltd
Sadler Road, Doddington
Road, Lincoln, Lincolnshire,
LN6 3RG, UK

Tel: 0522 684332

Littlewoods Organisation
J M Centre, 110 Old Hall
Street, Liverpool,
Merseyside, L70 1AB, UK

Tel: 051 235 2117

Lloyd Lifestyle Ltd
Langlands, Pallet Hill,
Penrith, Cumbria,
CA11 0BY, UK

Tel: 07684 83784

MZ Motorcycles GB Ltd
Royce Close, West
Portway, Andover,
Hampshire, SP10 3TS, UK

Tel: 0264 337443

Maidstone Wheel Centre
The Old Works, Dover
Street, Maidstone, Kent,
UK

Tel: 0622 720430

**Mead Competition
Motorcycles**
Fairlight Station Road,
Chambers Green, Pluckley,
Kent, TN27 ORL, UK

Tel: 0233 840323

Mead Speed
P.O.Box 536, Newport
Pagnell, Buckinghamshire,
MK16 8JN, UK

Tel: 0908 610311
Fax: 0908 617480

Suppliers of road and race
fairings, screens, classic racing
parts, brackets etc. Mail
order/export parts. As original
also carriers/grabrails. SAE for
catalogue.

Michael Brandon Ltd
15-17 Oliver Crescent,
Hawick, Roxburgh,
Borders, TD9 9BJ, UK

Tel: 0450 73333

Michelin Tyre PLC
The Edward Hyde Building,
38 Clarendon Road,
Watford, Hertfordshire,
WD1 1SX, UK

Tel: 0923 415000
Fax: 0923 415250

Michelin's massive investment in
research and development has
produced an extensive, high
quality range of motorcycle tyres
for all competition, sports,
touring and commuter bikes.

**Miles Engineering
Company**
Unit 4 Princes Works,
Princes Road, Teddington,
Middlesex, UK

Tel: 081 943 2022

**Mitsui Machinery Sales
UK Ltd**
Sopwith Drive, Brooklands,
Weybridge, Surrey,
KT13 0UZ, UK

Tel: 0932 358000
Fax: 0932 358030

Importers of Yamaha
motorcycles and mopeds.
Childrens machines and
specialised racing bikes for track
and off-road. Full range of
clothing and helmets also
imported.

Mobil Oil Co.Ltd
54 Victoria Street, London.
SW1E 6QB, UK

Tel: 071 828 9777

Morris Lubricants
Castle Foregate,
Shrewsbury, Shropshire,
SY1 2EL, UK

Tel: 0743 232200

**Motorcycle City Sales
Ltd**
149/151 Lynchford Road,
Farnborough, Hampshire,
GU14 6HD, UK

Tel: 0252 545086

**Motorcycle Electrical
Services**
Unit 10, Ladbrook Park Ind
Est, Millers Road,
Warwick, Warwickshire,
CV34 5AE, UK

Tel: 0926 499756

**Motorcycle Equipment
Ltd**
Watery Lane, Birmingham,
Midlands (W.), B9 4HE, UK

Contact: *R M Dennis*
Tel: 021 772 5944

Motorcycle Seatworks
366 Woodside Road,
Wyke, Bradford, Yorkshire
(W.), BD12 8HT, UK

Tel: 0274 604672

Manufacturers of every type of
motorcycle seat cover, specialists
in moto-cross. Appointment
always necessary.

**NGK Spark Plugs (UK)
Ltd**
7-8 Garrick Ind.Centre,
Hendon, London, NW9
6AQ, UK

Tel: 081 202 2151

**National Breakdown
Recovery**
PO Box 300, Leeds,
Yorkshire (W.), LS99 2LZ,
UK

Tel: 0532 393666

Oxford Products Ltd
Station Field Ind Est,
Kidlington, Oxford,
Oxfordshire, OX5 1JD, UK

Tel: 0865 841400

PSP Engineering Services
Units 5/5A Stalham
Rd.Ind.Est., Hoveton,
Norwich, Norfolk,
NR12 8DG, UK

Padgetts (Batley) Ltd
234 Bradford Road Batley,
Yorkshire (W.), WF17 6JD,
UK

Tel: 0942 478491

Phil Ayliff Products Ltd
25 Alliance Close,
Attleborough Fields,
Nuneaton, Warwickshire,
CV11 6SD, UK

Contact: *Trevor Ayliff*
Tel: 0203 343741
Fax: 0203 641247

Manufacturer of Dunlopad
sintered metal brake pads.

Phoenix Distribution Ltd
Units 9-10 Sneyd Ind Estate,
Burslem, Stoke or Trent,
Staffordshire, ST6 2DL, UK

Tel: 0782 838884

Piaggio Ltd
Unit 8, Ravensquay Centre,
Cray Avenue, Orpington,
Kent, BR5 4BQ, UK

Tel: 0689 898876
Fax: 0689 898862

Importers of scooters, including
the legendary Vespa range, also
Cosa, Sfera, Zip, Free, Quartz,
Skipper (125cc) and in 1994
Typhoon 50cc and 80cc sports
scooter.

Protectorl Ltd
Leekbrook Way, Cheadle
Road, Leek, Staffordshire,
ST13 7AS, UK

Tel: 0538 398377

R & R UK Ltd
Blackstone Road, Stukely
Meadow Ind. Estate,
Huntingdon,
Cambridgeshire, PE18 6EF,
UK

Tel: 0480 411146

RAC Motoring Services
114 Rochester Row,
London, SW1P 1Q, UK

Tel: 071 233 5711

Redcat Racing
1 North Road, Brighton,
Sussex (E.), BN1 1YA, UK

Tel: 0273 602944

Renold Chain Ltd
UK Sales Division,
Horninglow Road, Burton
on Trent, Staffordshire,
DE14 2PS, UK

Tel: 0283 512940

Renthal Ltd
Bredbury Park Way,
Bredbury, Stockport,
Cheshire, SK6 2SN, UK

Tel: 061 406 6399

Replica Fairings
Unit 3E Aston Ind.Estate,
Bulkington Road,
Bedworth, Warwickshire,
CV12 9DN, UK

Tel: 0203 311888
Fax: 0203 310262

The leading UK manufacturer
and supplier of motorcycle
bodywork. Massive range of
road, touring, sports, half, race
and twin headlamp fairings,
screens, mudguards, seat panels
and convertors, 1964-1994.

**Scott Leathers
International**
Unit 2, Stainton Ind.Estate,
Barnard Castle, Co.
Durham, DL12 8TZ, UK

Tel: 0833 31526

Serval Marketing Ltd
Serval House, Clifton Road,
Shefford, Bedfordshire,
SG17 5BZ, UK

Tel: 0462 815757

Shell Oils
Cobden House, Station
Road, Cheadle Hulme,
Cheshire, SK8 5AD, UK

Tel: 061 488 3134
Fax: 061 488 3153

Manufacturers and suppliers of
quality two and four-stroke
lubricants and speciality oils for
all motorcycle applications.
Available from Shell filling
stations and motorcycle dealers.

Silkolene Lubricants
Belper, Derbyshire,
DE5 1WF, UK

Tel: 0773 824151

Skoda (GB) Ltd
Bergen Way, North Lynn
Ind Estate, Kings Lynn,
Norfolk, PE30 2JH, UK

Tel: 0553 761176

**Sonic Communications
(Int) Ltd**
Communications Centre,
The Green, Castle
Bromwich, Birmingham,
Midlands (W.), B36 9AJ, UK

Tel: 021 749 4900

Sportex Gear
Units 4-8, Ripon Business
Park, Camphill Close,
Dalamires Lane, Ripon,
Yorkshire (N.), HG4 1QY,
UK

Tel: 0765 690800

**Stainless Engineering
Company**
Unit 9, Roman Way,
Longridge Road, Preston,
Lancashire, PR2 5BB, UK

Contact: *Andy Molnar*
Tel: 0772 700700

Replica stainles steel
components for British classic
machines. Replica hubs and
dural parts for competition use.
Available by mail order. Send
SAE for catalogue.

Staniforth Ltd
Church Street, Ecclesfield,
Sheffield, Yorkshire (S.),
S30 3WG, UK

Tel: 0742 462027

**Strathclyde Trading Co.
Ltd**
21 Nithsdale Street,
Glasgow, Strathclyde,
G41 2QA, UK

Tel: 041 423 2773

**Summerfield
Engineering Limited**
Cotes Park Industrial Estate,
Somercoates, Derbyshire,
DE55 4NJ, UK

Tel: 0773 833025

Supersprox
Station Works, Knucklas,
Knighton, Powys,
LD7 1PN, UK

Supreme Visors Ltd
Unit D5, Rofton Trading
Estate, Hooton Road,
Hooton, S.Wirral,
Merseyside, L66 7NS, UK

Tel: 051 327 7047

Suzuki/Information Dept.
PO Box 56, Tunbridge
Wells, Kent, TN1 2XY, UK

Tel: 0500 526262

Tec-Nick
92-98 Haltwhistle Road,
Western Industrial Area,
Chelmsford, Essex,
CM3 5ZF, UK

Tel: 0245 325985
Fax: 0245 325984

Manufacturers and suppliers of
electroplating kits, polishing and
abrasive materials to enthusiasts,
industry and trade. Callers to
our works by appointment only
please.

The Ultimate Source UK Ltd
Wham Hill, Croston Close Road, Ramsbottom, Bury, Lancashire, BL0 0RS, UK

Tel: 0706 823426

Thoroughbred Stainless
Unit 9, Potts Marsh Ind.Est., Westham, Pevensey, Sussex (E.), BN24 5NH, UK

Tel: 768564

Timeless Distributors Ltd
Ashford, Co.Wicklow, Eire

Tel: 010 353 404 40309

Top Tek International Ltd
3 Commerce Road, Stranraer, Dumfries & Galloway, DG9 7DX, UK

Tel: 0776 4421

Tregunna
6-8 Hatton Row, London, NW8 8PP, UK

Tel: 071 262 5678

Triple Cycles
228 Henley Road, Ilford, Essex, IG1 2TW, UK

Tel: 081 478 4807

Venhill Engineering Ltd
21 Ranmore Road, Dorking, Surrey, RH4 1HE, UK

Tel: 0306 885111
Fax: 0306-740535

Manufacturers of Nylocable and Featherlite control cables, Powerhose high performance stainless braided brake hoses. Distributors of Magura controls, Ariete and Buzzetti accessories and titanium and aluminium fasteners.

Vesty UK Ltd
Unit 2C, Merrow Business Ctr., Merrow Lane, Guildford, Surrey, GU4 7WA, UK

Tel: 0483 450560

Waddington MCA Ltd
131 Cottingham Road, Hull, Humberside (N.), HU5 2DH, UK

Tel: 0482 443101

Watsonian-Squire Ltd
Northwick Business Centre, Blockley, Gloucestershire, Gl56 9RF, UK

Tel: 0386 700907
Fax: 0386-700738

Major European manufacturer of sidecars and solo trailers. Full range of services include spares, fittings and technical advice. Factory showroom has examples of the largest range.

Westminster Communications Group
Cowley House, 9 Little College Street, London, SW1P 3SX, UK

Tel: 071 222 0666

Wheeler Racing
345 Ashley Road, Parkstone, Poole, Dorset, BH14 0AR, UK

Tel: 0202 742811

MUSEUMS

Automobilia Transport Museum
Billy Lane, Old Town, Wadsworth,Hebden Bridge, Yorkshire (W.), HX7 8RY, UK

Tel: 0422 844 775

Battlesbridge Motorcycle Museum
1 Maltings Road, Battlebridge, Essex, SS11 7RF, UK

Tel: 0268 769392

Birmingham Museum of Science and Industry
Newhall Street, Birmingham, Midlands (W.), B31 RZ, UK

Tel: 021 235 1661
Fax: 021 236 1766

The Museum's transport section includes motor cycles of both local and world-wide manufacture. Admission is free. The Museum is open daily 11am to 5pm.

Bristol Industrial Museum
Prince's Wharf, City Docks, Bristol, Avon, BS1 4RN, UK

Tel: 0272 251470

Industrial and transport history of Bristol area: Includes Douglas motorcycles and Quasar. Open Tues-Sun, 10-5 and bank holiday Mondays.

Brooklands Museum
The Clubhouse, Brooklands Road, Weybridge, Surrey, KT13 0QN, UK

Tel: 0932 857381

Caister Castle Motor Museum
Caister On Sea, Great Yarmouth, Norfolk, UK

Contact: Ald P R Hill
Tel: 057 284 251

Cheddar Motor & Transport Museum
Cheddar Gorge, Cheddar, Somerset, BS27 3QA, UK

Tel: 0934 742446

Combe Martin Motorcycle Collection
Cross Street, Combe Martin, Devonshire, EX34 0DH, UK

Contact: Terry McCulley
Tel: 0271 882346

The collection contains early and late British motorcycles displayed against a background of garage memorabilia. Open 10.00am to 5.00pm, end of May to end of September.

Cotswold Motor Museum
The Old Mill, Bourton on the Water, Gloucester-shire, GL54 2BY, UK

Contact: Mike Cavanagh
Tel: 0451 821255

A collection of thirty beautiful cars and motor cycles.

Design Museum
Butlers Wharf, Shad Thames, London, SE1 2YD, UK

Tel: 071 403 821255

East Anglia Transport Museum
Chapel Road, Carlton Colville, Lowestoft, Suffolk, NR33 8BL, UK

Contact: Mrs A Carr
Tel: 0986 798398

Easton Farm Park
Easton, Woodbridge, Suffolk, IP13 0EQ, UK

Tel: 0728 746475

Enfield & District Veteran Vehicle Trust
Whitewebb's Museum, Whitewebb's Road, Enfield, Middlesex, UK

Tel: 081 367 1898

Geeson Bros. Motorcycle Museum
Workshop, South Witham, Grantham, Lincolnshire, NG33 5PH, UK

Contact: G Geeson
Tel: 0572 7676280/767386

Eighty six British bikes dating back to 1913, most restored to new condition by ourselves. Refreshments available in workshop. Tel for 1994 open dates.

Grampian Transport Museum
Alford, Grampian, AB33 8AD, UK

Contact: M Ward
Tel: 09755 62292
Fax: 09755 62180

Museum of road and rail transport history. Motorcycles range from 1902 Beeston Humber to 1990 Norton Rotary. Open 10am-5pm daily, 27th March-30th Oct. Admission charge.

Haynes Motor Museum
Sparkford, Yeovil, Somerset, BA22 7LH, UK

Contact: M Penn
Tel: 0963 40804
Fax: 0963 40877

One of the largest motor museums in the country with a good selection of rare and interesting motorcycles: classic, veteran, vintage and American motorcars.

Historic Vehicles Collection
63-67 High Street, Rolvenden, Cranbrook, Kent, TN17 4LP, UK

Contact: C.M.Booth
Tel: 0580 241234

Museum specialising in Morgan 3-Wheelers, plus other cars, motorcycles, cycles, models, toys, automobilia. Open Monday to Saturday 10am-6pm, and occasionally on Sundays.

Jersey Motor Museum
St Peter's Village, Jersey, CI, JE3 7AG, UK

Tel: 0534 482966

Lakeland Motor Museum
Holker Hall, Cark in Cartmel, Cumbria, LA11 7PL, UK

Tel: 05395 58509

Open 1st April to 31st October, Sunday to Friday, (closed saturdays) 10.30am-5.00pm. Last admissions 4.30pm.

Lambretta Museum
Kesterfield, Northlew, Okehampton, Devonshire, EX20 3PN, UK

Tel: 0409 221488

Manx Motor Museum
Crosby, Isle of Man, UK

Contact: R Evans
Tel: 0624 851236

Midland Motor Museum
Stourbridge Road, Bridgnorth, Shropshire, WV15 6DT, UK

Tel: 0746 761761

The museum houses many of the world's fastest and most successful cars and motorcycles. Over 100 vehicles on display in the converted stables of Stanmore Hall. Curator: Mike Barker.

Murrays Motorcycle Museum
Bungalow Corner TT Course, Snaefell Mountain, Isle of Man, UK

Contact: C Murray
Tel: 062 486 719

Museum of British Road Transport
St Agnes Lane, Hales Street, Coventry, Midlands (W.), CV1 1PN, UK

Tel: 0203 832425

Museum of Science & Industry
Newhall Street, Birmingham, Midlands (W.), B3 1RZ, UK

Contact: J Andrew
Tel: 021 235 1661

Museum of Transport
Kelvin Hall, 1 Bunhouse Road, Glasgow, Strathclyde, G3 8DP, UK

Tel: 041 357 3929
Fax: 041 357 5849

A museum devoted to the history of transport, including a large display of motor cycles from the earliest days.

Myreton Motor Museum
Aberlady, Lothian (E.), EH32 OPZ, UK

Contact: M Mutch
Tel: 0875 870288

A large collection of motor cycles from 1902, including Brooklands Charterlea and 1912 Henderson. Cars, commercials, cycles, British military vehicles, advertising, and ephemera.

National Motor Museum
Beaulieu, Brockenhurst, Hampshire, SO42 7ZN, UK

Contact: Annice Collett
Tel: 0590 612345
Fax: 0590 612655

Open 364 days a year from 10am. Over 50 motorcycles and engines on display. We also hold handbooks, road tests etc for many makes.

National Motorcycle Museum
Coventry Road, Bickenhall, Solihull, Midlands (W.), B92 OEJ, UK

Tel: 06755 3311

Oswestry & DMC
Sherbrooke, Middleton Road, Oswestry, Shropshire, UK

Contact: Mr W H Jones
Tel: 0691 656471

Oswestry Bicycle Museum
Oswald Road, Oswestry, Shropshire, SY11 1RE, UK

Contact: David Higman
Tel: 0691 671749

Oswestry Bicycle Museum opens daily displaying the history and development of the bicycle with 100 exhibits. Displays include attachment engines and mopeds. Large Dunlop archive.

Royal Museum of Scotland
Dept.of Science & Technology, Chambers Street, Edinburgh, Lothian, EH1 1JF, UK

Tel: 031 225 7534

Sammy Miller Museum
Gore Road, New Milton, Hampshire, UK

Tel: 0425 619696

Science Museum
Exhibition Road, South Kensington, London, SW7 2DD, UK

Tel: 071 589 3456

Open: 10.00-18.00 Mon-Sat, 11.30-18.00 Sun. The large Land Transport Gallery includes motorcycles and engines from 1894 onwards. Archives of British motorcycle companies.

Stanford Hall Motorcycle Museum
Lutterworth, Leicestershire, LE17 6DH, UK

Tel: 0788 860250

Outstanding collection of vintage and rare racing motorcycles dating from 1915 to present day. Open Easter - end September, Saturdays, Sundays and Bank Holiday Mondays, 2.30-5.30pm.

Totnes Motor Museum
Steamer Quay, Totnes, Devonshire, TQ9 5AL, UK

Tel: 0803 862777

Velocette Heritage
The Rhodes Collection, Fellside, Cole Lane, Borrowash, Derbyshire, DE7 3GN, UK

Tel: 0332 673658

West Wycombe Motor Museum
Chorley Road, West Wycombe, Buckinghamshire, UK

Tel: 0494 443329

REPAIRS

A. Gagg & Sons
106 Alfreton Road,
Nottingham,
Nottinghamshire,
NG7 3NS, UK

Tel: 0602 786288

Armstrong Engineering
176 Westgate Road,
Newcastle upon Tyne, Tyne
& Wear, NE4 6AL, UK

Tel: 091 261 4579

Ashtech Limited
Victory Works, Garden
Field, Wyke, Bradford,
Yorkshire (W.),
BD12 9NH, UK

Tel: 0274 679071

B Perry
9 Pond Walk, Cranham,
Essex, RM14 3YH, UK

Bob Jackson
Rose & Crown Works,
Stricklandgate, Kendal,
Cumbria, UK

Tel: 0539 720582

C J Wilson
23-25 West Main Street,
Uphall, Broxburn, Lothian,
UK

Tel: 0506 856751

CES Ltd
19 Blacksmith Close, North
Springfield, Chelmsford,
Essex, CM1 5SY, UK

Tel: 0245 469212

Carl Rosner Ltd
Station Approach,
Sanderstead Road, South
Croydon, Surrey, CR2
0PL, UK

Contact: Carl Rosner
Tel: 081 657 0121
Fax: 081 651 0596

Established 25 years.
Comprehensive spares stock for
Commando, Triumph 650/750.
Trident/R-3. Full workshop
facilities, including restorations.
Dealers for Hinckley Triumphs,
Norton Rotary and Indian
Enfield.

Chris Applebee
471 Rayleigh Road,
Benfleet, Essex, UK

Coleman Body Repairs
40C Humber Avenue,
Coventry, Midlands (W.),
CV3 1AY, UK

Tel: 0203 458494

**Coulson Engineering
Services Ltd**
Roselawn Farm, 110 Main
Road, Broomfield,
Chelmsford, Essex,
CM1 5AG, UK

Tel: 0245 73696

Cylinder Head Shop
28-30 Broadway Court,
Wimbledon, London,
SW19 1RG, UK

Contact: Len Paterson
Tel: 081 946 2434
Fax: 081 879 3833

Specialists in leadfree head
conversions, also "STD" and
performance seat cutting and
valve guide lining - special and
STD guides made/fitted utilising
latest technology equipment.

Daytona
42-48 Windmill Hill, Ruislip
Manor, Middlesex,
HA4 8PT, UK

Tel: 0895 675511
Fax: 0895 630654

Dealership for Kawasaki,
Triumph and Ducati
motorcycles. Also main parts
stockist for Kawasaki.

Esoteric
Unit 2, Greenstar Buildings,
363A Lytham Road,
Blackpool, Lancashire,
FY4 1EA, UK

FTW Motorcycles
178 Penistone Road North,
Sheffield, Yorkshire (S.), UK

Tel: 0742 336269

Fine Thompson Ltd
Felspar Road, Amington,
Tamworth, Staffordshire,
B77 4DP, UK

G&D Motorcycles
Clothallbury Ind Est,
Buntingford Road, Baldock,
Hertfordshire, SG7 6RG,
UK

Tel: 0462 79729

Gowrings
245 Finchampstead Road,
Wokingham, Berkshire, UK

Tel: 0734 770350

Graham Engineering
Roebuck Lane, West
Bromwich, Birmingham,
Midlands (W.), B70 6QP,
UK

Tel: 021 525 3133

Hobbs Sport Earls Ltd
4D Brent Mill Ind.Estate,
South Brent, Devonshire,
TQ10 9YT, UK

Tel: 0364 73956
Fax: 0364 73957

Hughie Hancox
21 Bayton Road, Exhall,
Coventry, Midlands (W.),
UK

Tel: 0203 368038

**Independent Ignition
Supplies**
Myrtle Street, Appledore,
Bideford, Devonshire,
EX39 1PH, UK

Tel: 0237 475986

J.W. Tennant-Eyles
Barcote Manor, Buckland,
Nr.Faringdon, Oxfordshire,
SN7 8PP, UK

Tel: 036 787 330

KAIS
Punchbowl Garage,
Atherton, Manchester,
Manchester, Gt., UK

Tel: 0942 896366

Kennedy Motors
Ford Farm, Walsh Road
West, Southam,
Warwickshire, UK

Tel: 092681 7374

Kinetic Art Paintshop
Unit 25,Riverside Business
Pk., Lyon Road, Merton,
London, SW19 2RL, UK

Tel: 081 544 9376/9377
Fax: 081 544 9376/9377

Motorcycle accident repair
centre. Modification, fabrication
and custom paintwork to bikes
and crash helmets.

Leslie Griffiths Motors
10A Ewenny Road,
Bridgend, Glamorgan, Mid-,
UK

Tel: 0656 661131/2

M R Holland
P.O.Box 53, Spalding,
Lincolnshire, PE11 3UX, UK

Tel: 0775 766455

Mercer Skilled Crafts Ltd
Springfield Works,
Moorside, Cleckheaton,
Yorkshire (W.), BD19 6JT,
UK

Tel: 0274 872861

Merlin Motorsport
Castle Combe Circuit,
Chippenham, Wiltshire,
SN14 7EX, UK

Tel: 0249 782101

NDT Motorcycles
Banks Bldgs., Front Street,
New Herrington, Houghton
le Spring, Tyne & Wear,
DH4 7AU, UK

Tel: 091 584 8881

Nova
8 Horseshoe Yard,
Crowland, Lincolnshire,
PE6 0BJ, UK

Tel: 0733 210082

P & M Motorcycles
8 Set Star Estate, Transport
Avenue, Brentford,
Middlesex, TW8 9HF, UK

Tel: 081 847 1711
Fax: 081 758 1403

Repairs and service to most
makes of motorcycles old and
new. Full engineering and
welding facilities. Collection of
machines if required.

Performoto
Unit 6, Enterprise Row,
Rangemoor Industrial Estate,
Tottenham, London,
N15 4NG, UK

Tel: 081 880 3420
Fax: 081 880 9330

Motorcycle engineers
specialising in chassis
modifications, thread repairs,
frame straightening, steel/alloy
welding, fabrications and special
building. Full engineering
facilities with friendly,
enthusiastic staff. Open 9.30-6.30
weekdays and 10-4 Saturdays.

Picador Engineering Ltd
Foxhills Industrial Estate,
Scunthorpe, Humberside
(S.), DN15 8QJ, UK

Priest Motor Bodies
6 Viaduct Road,
Broadheath, Altrincham,
Cheshire, WA14 5DU, UK

Tel: 061 928 3210

Readspeed
79 Station Lane,
Featherstone, Pontefract,
Yorkshire (W.), UK

Rockhall Auto Electrics
Valley Road, Hayfield,
Stockport, Cheshire,
SK12 5LP, UK

Tel: 0663 742539

Roger Titchmarsh Racing
Primrose Cottage, East
Cottingwith, Yorkshire,
YO4 4TH, UK

Tel: 0759 318562

Route 66
204A Walsall Wood Road,
Aldridge, Midlands (W.),
WS9 8HB, UK

Tel: 0922 59398

SRM Engineering
Unit 22, Glan-yr-Afon Ind
Est, Aberystwyth, Dyfed,
SY23 3JQ, UK

Tel: 0970 627771/2

Sam Pearce & Son
Unit No5 Stanley Lane,
Bridgnorth, Shropshire,
WV16 4SF, UK

Tel: 0746 762743

Scott Oilers
106 Clober Road, Mingavie,
Glasgow, Strathclyde,
G62 7SS, UK

Tel: 041 956 6225

Shirlaws Motorcycles
92 Crown Street,
Aberdeen, Grampian,
AB1 2HJ, UK

Contact: *Roy Shirlaw*
Tel: 0224 584855

Mail order on bikes by BMW,
Ducati, Triumph, Yamaha,
Suzuki, Honda, Vespa. Parts,
clothing & helmets, specialist
repairs, training. Motorcycle
world all under one roof.

Stan Stephens
6 Portobello Parade,
Fawkham Road, West
Kingsdown, Kent, UK

Taylor Racing
23-25 Station Hill,
Chippenham, Wiltshire, UK

Tel: 0249 657575/6

**Tommy Robb
(Motorcycles) Ltd**
240 Manchester Road,
Warrington, Cheshire,
WA1 3BE, UK

Tel: 0925 56528
Fax: 0925 234518

Main agents for Honda,
Triumph, Kawasaki, Yamaha,
Piaggio motorcycles. We offer a
vast range of accessories:
clothing, helmets, boots, gloves,
etc.; plus advice on training,
insurance, finance etc. Our
service department is the best for
repairs and MOTs.

Tony Holmshaw
5-6 Acre Road, March,
Cambridgeshire, PE15 9JD,
UK

Tel: 0354 56345

Trident Engineering
343A Rayners Lane, Pinner,
Middlesex, HA5 5EN, UK

Tel: 081 868 1476

Repairs, parts, tuning and race
preparation. Full workshop
facilities including milling,
turning, cylinder head and brake
disc skimming, helicoils etc.
Backed by T.T. winning
experience.

Twiggers Motorcycles
30 Nottingham Road,
Loughborough,
Leicestershire, LE11 1EU,
UK

Tel: 0509 263967

Honda motorcycles stocked in
Leicestershire (off Junction
23-M1). Twiggers Motorcycles
for new Honda sales, good used
motorcycles bought and sold.
Service work in Honda
approved workshop, repairs,
insurance and spares for most
makes. MOTs, tyres supplied,
fitted, balanced and repaired.

Wire Wheels Services
26 Bowesfield Lane,
Stockton-On-Tees,
Cleveland, TS18 3ER, UK

Tel: 0642 604768

Section 10

RESTORATION

A E Pople
52 Henley Drive, Frimley
Green, Surrey, GU16 6NF,
UK

Tel: 0252 835353

APS Sales
Steinkirk Building,
93 Dunkirk Road, Lincoln,
Lincolnshire, LN1 3UJ, UK

Contact: S Farrant
Tel: 0522 542366

Restoration, fasteners, nuts,
bolts, Allen screws in stainless
chrome, cadmium, in BSCY, BSF,
UNF, WHIT/UNC metric.
Polished stainless to order.
Chrome, polished hexagon &
heads. Mail order available.
Open Mon, Tues & Friday
10.30am-5.00pm. and late
Thursday 10.30am-8.00pm.
Closed Wednesday.

ACE Electroplaters
64 Danby Walk, Upper
Accomodation Road, Leeds,
Yorkshire (W.), LS9 8JE,
UK

Tel: 0532 480972

Ace Classics London
101-103 St Mildreds Road,
Lee, London, SE12 0RL,
UK

Tel: 081 698 4273

Alloy Polishing Service
9 Teinkirk Building,
93 Dunkirk Road,
Lincoln, Lincolnshire, LN1
3UJ, UK

Contact: M.Farrant
Tel: 0522 542366

Restoration sale of metal
polishing materials (as used in
our own workshop) for D.I.Y.
enthusiasts. Backed by chrome
plating and metal polishing
service. For a mail order list send
a s.a.e. Open Mon, Tues & Fri.
10.30am - 5.00pm. Closed
Wednesdays. Late night
Thursday 10.30am-8.00pm.

Allwick Patterns
Seager Road, Oare,
Faversham, Kent,
ME13 7TL, UK

Tel: 0795 532580

Ariel
The Old Town Maltings,
Broad Street, Bungay,
Suffolk, NR35 1EE, UK

Tel: 0986 894798

Bob Price Classic Bikes
14 The Crescent, Leek,
Staffordshire, ST13 6HD,
UK

BSA A50/65 unit twin specialist.
New and used parts.
Restorations, engine rebuilds.
Write or phone Bob on 0538
385939 daytime 1pm-7pm.

Carl Rosner Ltd
Station Approach,
Sanderstead Road, South
Croydon, Surrey, CR2
0PL, UK

Contact: Carl Rosner
Tel: 081 657 0121
Fax: 081 651 0596

Established 25 years.
Comprehensive spares stock for
Commando, Triumph 650/750.
Trident/R-3. Full workshop
facilities, including restorations.
Dealers for Hinckley Triumphs,
Norton Rotary and Indian
Enfield.

**Charnwood Classic
Restorations**
107 Central Road,
Hugglescote, Coalville,
Leicestershire, LE6 2FL, UK

Tel: 0530 832257

Chris Elderfield
25 Church Street,
Goldalming, Surrey,
GU7 1EL, UK

Tel: 0483 427940

**Classic & Custom
Motorcycles**
125 Chester Road,
Northwich, Cheshire,
CW8 4AA, UK

Tel: 0606 77262
Fax: 0606 77262

Pitted forks? Forks ground and
hardchromed to the highest
standards. Nine day turnround;
Securicor collection and delivery
guaranteed better than new!
Faxed information sheet
available. Callers by prior
arrangement only.

Dan Force Restorations
58 Dishley Street,
Leominster, Hereford &
Worcester, HR6 8NY, UK

Tel: 0568 611837

David Newman
Farnborough Way,
Farnborough, Kent,
BR6 7DH, UK

Tel: 0689 857109

Dragonfly Motorcycles
The Old Town Maltings,
Broad Street, Bungay,
Suffolk, NR35 1EE, UK

Tel: 0986 894798

**Fareham Electroplating
Co. Ltd**
Unit 25, Fort Fareham,
Newgate Lane, Fareham,
Hampshire, PO14 1AH, UK

Contact: Mrs P Boulin
Tel: 0329-288831

Enquiries, personal callers, mail
orders welcomed (fully insured
return postage). Friendly, expert
restoration service.

G&D Motorcycles
Clothallbury Ind Est,
Buntingford Road, Baldock,
Hertfordshire, SG7 6RG,
UK

Tel: 0462 79729

GU Products
Unit 9, Abersychan
Ind.Estate, Pontypool,
Gwent, NP4 7BA, UK

Tel: 0495 774953

Gatefield Classic Motorcycles
Unit 2, Wellington Road,
Ashton under Lyme,
Lancashire, UK

Tel: 061 343 4774

George Hopwood Motorcycles
112 Wren Road, Sidcup,
Kent, DA14 4NF, UK

Contact: George Hopwood
Tel: 081 300 9573

Triumph specialist. Engine
overhauls. Total restorations.
T120 Thruxton parts - exhausts
pipes, camshafts seats, fairings,
etc. Full tuning services. Free
advice and discounts to TOMCC
members.

Joe Francis Motors Ltd
340 Footscray Road, New
Eltham, London, SE9 2ED,
UK

Tel: 081 850 1373
Fax: 081 859 5617

John Mossey Restorations
Unit 11 Cherry Tree Farm,
Cambridge Road,
Melbourne,
Cambridgeshire, SG8 6EX,
UK

Tel: 0763 260096

Kirby Rowbotham
58 Arch Street, Rugeley,
Staffordshire, WS15 1DL,
UK

Tel: 0889 584758

LB Restorations
Billinghurst, Sussex (W.),
UK

Tel: 0403 783478

Lewis & Templeton
Brinklow Service Station, 26
Coventry Road, Brinklow,
Rugby, Warwickshire,
CV23 ONE, UK

Tel: 0788 833330

Lightning Bolt Co.
P.O.Box 69, Rochester,
Kent, ME1 1XX, UK

Contact: Chris Bloomfield
Tel: 0634 271276
Fax: 0634 271276

The stainless steel fasteners
(nuts, bolts etc) specialist for
Harleys. Screwkits and
individual UNF and UNC
fasteners. Specials e.g headbolts.
Free price list and advice.

Martin Arscott
Warren Cottage, Tinkers
Lane, Tring, Hertfordshire,
HP23 6JB, UK

Tel: 0442 862966

Velocette spares, repairs and
servicing of all components;
threads reclaimed, brake drums
skimmmed on wheels, valve
seats, complete machine bulding,
O.M.C. a speciality.

Norman White Norton
Thruxton Circuit, Andover,
Hampshire, UK

Tel: 0264 773326

Nourish Racing Engine Company
13 Manor Lane, Langham,
Oakham, Leicestershire,
LE15 7JL, UK

Tel: 0572 722712

Phil Carter Motorcycles
270 London Road,
Northwich, Cheshire, UK

Tel: 0606 47627

Phoenix Supplies
PO Box 207, Lincoln,
Lincolnshire, LN6 8US, UK

Tel: 0522 696988

Pollards Motorcycles
The Garage, Clarence
Street, Dinnington,
Sheffield, Yorkshire (S.),
S31 7HA, UK

Tel: 0909 563310

Power Torque Engineering Ltd
Herald Way, Binley,
Coventry, Midlands (W.),
CV3 2RQ, UK

Tel: 0203 635757

Precision Engineering Services
P O Box 36, Saffron
Walden, Essex, CB11 4QE,
UK

Tel: 0799 528388

Spares and service for Amal,
Monobloc and Pre-Monobloc
carburettors. Worn throttle slides
repaired, carburettors
overhauled and restored. Parts
available by post.

Priest Motor Bodies
6 Viaduct Road,
Broadheath, Altrincham,
Cheshire, WA14 5DU, UK

Tel: 061 928 3210

RCM
Unit 15, Argyle Way, Ely,
Cardiff, Glamorgan (S.), UK

Tel: 0222 593122

R. K. Leighton
Unit 5, Gunsmith House,
50-54 Price Street,
Birmingham, Midlands (W.),
B4 6JZ, UK

Tel: 021 359 0514

RJM
Ram Yard Estate, High
Street, Arlesey,
Bedfordshire, SG15 6SW,
UK

Tel: 0462 732862

Full restorations, spare parts,
repairs and all restoration
services. For classic British
motorcyclists specialising in
BSA, Triumph and Norton.
Open late seven days a week.

Racespec
Yauncos Cottage, Tillers
Green, Dymock,
Gloucestershire, GL18 2AP,
UK

Tel: 0531 890250

Richard Hughes
90 Redacre Road, Sutton
Coldfield, Midlands (W.),
B73 5EE, UK

Tel: 021 354 7524

Richards Brothers
56 Clive Road, Canton,
Cardiff, Glamorgan (S.),
CF5 1HG, UK

Tel: 0222 229945
Fax: 0222 641051

Specialist wheelbuilders since
1937. Shot blasting and stove
enamelling of frames etc.
Stainless steel and British made
chrome rims in stock.
Nationwide collection service.

Robin James Engineering Services
Clinton Road, Leominster,
Hereford & Worcester,
HR6 ORJ, UK

Tel: 0568 612800

Motorcycle restoration
engineers: Mon-Fri, 8.00-1.00 -
2.00-5.00. No problem too big or
too small! All work undertaken,
including painting.

Roger Titchmarsh Racing
Primrose Cottage, East
Cottingwith, Yorkshire,
YO4 4TH, UK

Tel: 0759 318562

Rossendale Wheels
175 Burnley Road,
Rawtenstall, Lancashire, UK

Tel: 0706 226127

SERVICES

SC Engineering
Units 48/53 Hardwick Ind
Est., Bury St Edmunds,
Suffolk, IP33 2QH, UK

Tel: 0284 703615

Portable pressure shot blasters.
Pressure pots for cabinets. Fly
wheel trueing centres.
Specialised parts machined to
order. Alloy welding. Frame
repairs and blast cleaning.

Scientific Coatings
13 Woulton Close,
Ashton-In-Makerfield,
Manchester, Gt., UK

Sevenhills Motorcross
Sevenhills House, Ingham,
Bury St Edmunds, Suffolk,
UK

Tel: 0284 728018

Steve Bullock
Unit 20, East Coast Tdg.
Est., West Lynn, Kings
Lynn, Norfolk, UK

Tel: 0553 771466

**Steve Tonkin
Restorations**
North Road Garage, Lower
North Road, Carnforth,
Lancashire, LA5 9LJ, UK

Tel: 0524 733222

TT Restorations
Unit 34A, Warwick Street,
Coventry, Midlands (W.),
CV5 6ET, UK

Tel: 0203 713404

TDR Restorations
80 Dartford Road,
Dartford, Kent, DA1 3ER,
UK

Tel: 0322 229232
Fax: 0322 278752

Engine rebuilds, gearbox
rebuilds, all mechanical work,
tanks sprayed, stove enamelling,
powder coating, bead blasting,
clocks rebuilt, welding,
servicing, repairs, wheel rebuilds
and full restorations.

JW Tennant-Eyles
Barcote Manor, Buckland,
Faringdon, Oxfordshire,
SN7 8PP, UK

Tel: 036787 330
Fax: 036787 416

The Finishing Touch
26 East Hanningfield Rd,
Rettendon, Chelmsford,
Essex, CM3 8EQ, UK

Tel: 0243 400918

Classic motor cycle paintwork,
to restore rust, dents, pitting.
Tank repairs, chroming, colour
matching cellulose - two pack.
Paint lining - wheels painted and
lined. Complete motorcycles or
parts restored.

**The Goodman
Engineering Co. Ltd**
Westward, Buckle St,
Honeybourne, Evesham,
Hereford & Worcester,
WR11 5QQ, UK

Tel: 0386 832090
Fax: 0386 831614

The Paint Studio
95 Station Road, Ilkeston,
Derbyshire, DE7 5LL, UK

Tel: 0602 322290

VMCC Transfer Service
Arosfa, Cwmpennar,
Mountain Ash, Glamorgan,
Mid-, CF45 4DL, UK

VMCC Regalia
112 Fairfield Crescent,
Newhall, Swadlincote,
Derbyshire, DE11 OTB,
UK

Tel: 0283 224841

Weeden Restoration
Unit 4, Atlas Court,
Hermitage Ind. Estate,
Coalville, Leicestershire,
UK

Tel: 0530 811118

A J Whitehouse
PO Box 35, Shirley,
Solihull, Midlands (W.),
B90 3QD, UK

A44 Motorcycles
3 Etnam Street, Leominster,
Hereford & Worcester, UK

Tel: 0568 612564

Large selection of quality used
machines; P/X & H/P terms;
bikes bought for cash; helmets
and clothing; computerised parts
service; Yokohama tyres, etc.

ALAPAT
Unit 28B, Escott Works,
Rome Street, Carlisle,
Cumbria, CA2 5LE, UK

Tel: 0228 819324

Chrome, nickel, copper, zinc
plating. Wheel building, brake
drum/disc skimming, milling,
turning, helicoiling, fork
straightening. Vintage
repairs/restoration. Metal
polishing. All above done on
premises.

**Abacus Investments
Europe Ltd**
St Ronans Research, 191
Park View, Whitley Bay,
Tyne & Wear, NE26 3RD,
UK

**Advanced Motorcycle
Systems**
22 Tintern Avenue,
Kingsbury, Middlesex,
NW9 ORJ, UK

Contact: Mark Wonnacott
Tel: 081 204 3904

Alarm systems for all types of
motorcycles from paging, remote
and manual. Weekend
installations by appointment,
cheapest prices guaranteed,
discount to clubs.

Anglo-Scot Abrasives
5 Bolton Road, Ashton in
Makerfield, Manchester,
Gt., WN4 8AA, UK

Tel: 0942 270729

Armstrong Engineering
176 Westgate Road,
Newcastle upon Tyne, Tyne
& Wear, NE4 6AL, UK

Tel: 091 261 4579

Autobikers Ltd
116 Henley Road,
Caversham, Reading,
Berkshire, RG4 ODH, UK

Tel: 0734 474155

Automobile Association
Fanum House, Basingstoke,
Hampshire, RG21 2EA, UK

Tel: 0256 20123

BD Engineering
Unit G1, Newington
Industrial Estate,
Newington, Kent, UK

Tel: 0795 843980

BMF
35 Gellantly Place, Brechin,
Angus, Tayside, DD9 6BS,
Scotland

Contact: Stewart Mowatt
Tel: 0356 623981

BMF
13 South Terrace, Sowerby,
Thirsk, Yorkshire (N.),
YO7 1RH, UK

Contact: John Carrington
Tel: 0845 524264

BSA Company Limited
Units 98 & 99, Northwick
Park Business Centre,
Blockley, Gloucestershire,
GL56 9RF, UK

Tel: 0386 700753
Fax: 0386 700435

BSS Motorcycles
7 Martins Lane, Wallasey,
Merseyside, UK

Tel: 051 638 3736

Baker Engineering
Paramount Ind.Est.,
Sandown Road, Watford,
Hertfordshire, WD2 4XA,
UK

Tel: 0923 229309

Barber Engineering
Bunwell Road, Besthorpe,
Attleborough, Norfolk,
NR17 2NZ, UK

Contact: *Denny Barber*
Tel: 0953 452422

Classic frames. Manx G50,
Seeley, Aemacchi, Ducati.
Speedway Rotax MK1,
Speedway Rudge, Grass Track.
Other frames to pattern or
drawing. SAE for enquiries.

Barclay Motorcycles
179 Ashley Road,
Bournemouth, Dorset, UK

Tel: 0202 394367

**Barclays Business
Systems**
644 Wimbourne Road,
Bournemouth, Dorset,
BH9 2EH, UK

Contact: *Bill Barclay*
Tel: 0202 546464
Fax: 0202 515880

The Biketech datafinder gives a
wealth of technical and reference
information for most models
manufactured in the last 20 years.

**Bradford Ignition
Services**
Unit 6, Thorncliffe Ind.Est.,
Thorncliffe Road, Bradford,
Yorkshire (W.), BD8 7DD,
UK

Tel: 0274 546268

Bassets
12 Howard Place, Stoke on
Trent, Staffordshire, UK

Tel: 0782 212890

Bath Road Motorcycles
379-387 Bath Road,
Brislington, Bristol, Avon,
UK

Tel: 0272 711447

**Bickers Anglia (Access)
Ltd**
Units 7 & 8, Farthing Rd.
Est., Sproughton Road,
Ipswich, Suffolk, IP1 5AP,
UK

Tel: 0473 745131

Importers and distributors of
motorcycle spares, to the trade
only. Specialising in equipment
equivalent to O.E.M. standard.
For more details contact your
local dealer.

Bike Monsters
12 Newfields Business Park,
Hatch Pond Road, Poole,
Dorset, BH17 7NF, UK

Contact: *Nick Holmes*
Tel: 0202 735856
Fax: 0202 660284

Importers of Bike to Bike
Intercomms call "The
Communicator" & helmet
speakers "Bass Monsters".
Any enquiries please call.

Bike-Teck
Limes Garages, A22
Eastbourne Road, Blindley
Heath, Surrey, RH7 6JJ, UK

Bob Jackson
Rose & Crown Works,
Stricklandgate, Kendal,
Cumbria, UK

Tel: 0539 720582

Bob Newby
1 Arnills Way, Kilsby,
Rugby, Warwickshire,
CV23 8UY, UK

Tel: 0788 822816

**Brands Hatch Leisure
PLC**
Fawkham, Longfield, Kent,
DA3 8NG, UK

Tel: 0474 872331

**British Motorcycle
Engineering**
A4 Penarth Dock, Penarth,
Glamorgan (S.), CF6 1LR,
UK

Tel: 0222 708132

Bryan Goss Motorcycles
8 Yeo Valley Business
Centre, Stoford, Yeovil,
Somerset, BA22 9US, UK

Tel: 0935 72424

C & I Threading
PO Box YR8, Leeds,
Yorkshire (W.), LS9 9HX,
UK

Tel: 0937 582435

C J Wilson
23-25 West Main Street,
Uphall, Broxburn, Lothian,
UK

Tel: 0506 856751

C Wylde & Son Ltd
103 Roundhey Road, Leeds,
Yorkshire (W.), LS8 5AJ,
UK

Tel: 0532 491856

CES Ltd
19 Blacksmith Close, North
Springfield, Chelmsford,
Essex, CM1 5SY, UK

Tel: 0245 469212

**Cap Nationwide Motor
Res. Ltd**
CAP House, Carleton
Road, Skipton, Yorkshire
(N.), BD23 2BE, UK

Tel: 0756 700666

**Caravan Camping &
Leisure**
42 Cromer Road, West
Runton, Norfolk, NR27
9AD, UK

Tel: 0263 837482

Camping and outdoor (mail
order) * tents * sleeping bags *
mats * stoves * lights * cookware
* accessories * spares * repair
services * plus much more! Free
mail order lists.

Castleford M/C
Unit 25 Raglan Works,
Mehtley Road, Castleford,
Yorkshire (W.), UK

Tel: 0977 515427

Cat Motorcycles
99-105 Upper Stone Street,
Maidstone, Kent, UK

Tel: 0622 685959

Classic Brake Services
PO Box 5, Whaley Bridge,
Stockport, Manchester, Gt.,
SK12 7LL, UK

Contact: *Alan Campbell*
Tel: 0663 732940

All types of motorcycle brake
shoes relined, from road type
vintage to classic racing.
Standard thickness or oversize
linings. Friction damper discs
made to order.

Clinton Enterprises
Spring Cottage, Main
Street, Cadeby, Nuneaton,
Warwickshire, CV13 0AX,
UK

Tel: 0455 290976

Colin French
Unit 6, Kingsley,
Hampshire, GU35 9LY, UK

Tel: 0420 489676

Colin Gregory M/Cs
835 Mansfield Road,
Nottingham,
Nottinghamshire, UK

Tel: 0602 260155

Collins Tyre Service
468 London Road,
Portsmouth, Hampshire,
UK

Tel: 0705 694741

**Cosmopolitan Motors
Limited**
73/75 Camberwell Road,
London, SE5 0EZ, UK

Tel: 071 703 2271

Crafty Toys
8a Douglas Road,
Maidstone, Kent,
ME16 8ES, UK

Tel: 0622 755293

Crowmarsh Classic Motorcycles
Unit 3, Honey Farm, Preston, Crowmarsh, Oxfordshire, OX10 6SL, UK

Tel: 0491 25869

Crusader Bike Tours Ltd
68-69 The Mint, Rye, Sussex (E.), UK

Tel: 0797 224640

Custom Fasteners Ltd
Unit 64, Mochdre Ind.Estate, Newtown, Powys, SY16 4LE, UK

Tel: 0686 629666
Fax: 0686 622620

Mail order accessories - nuts and bolts in chrome, stainless, zinc and nickel plus a huge range of custom parts. Open Monday to Saturday 9.00-5.00.

Cylinder Head Shop
28-30 Broadway Court, Wimbledon, London, SW19 1RG, UK

Contact: Len Paterson
Tel: 081 946 2434
Fax: 081 879 3833

Specialists in leadfree head conversions, also "STD" and performance seat cutting and valve guide lining - special and STD guides made/fitted utilising latest technology equipment.

DJF Metal Polishing
249 Ladypool Road, Sparkbrook, Birmingham, Midlands (W.), B12 8LF, UK

Tel: 021 446 4215

David Earnshaw
71 Quarry Hill Road, Wath upon Dearne, Rotherham, Yorkshire (S.), S63 7SZ, UK

Dick Weekes
135 Newton Road, Newton, Swansea, Glamorgan (W.), SA3 4ST, UK

Tel: 0792 367635

Calthorpe motorcycles - 1908 to 1939 - information service data/specifications. Machine authentication. VMCC marque specialist service free to UMCC members.

Dorset Dirt Bikes
345 Ashley Road, Parkstone, Poole, Dorset, UK

Tel: 0202 721500

Dragonfly Motorcycles
The Old Town Maltings, Broad Street, Bungay, Suffolk, NR35 1EE, UK

Tel: 0986 894798

Eddy Grew
17 Brookdale, Hinckley, Leicestershire, LE10 0NX, UK

Tel: 0455 637048

Epoxy Powder Coatings Ltd
215 Tyburn Road, Birmingham, Midlands (W.), B24 8NB, UK

Tel: 021 328 2145

Exports-Imports
16 Begonia Drive, Burbage, Leicestershire, LE10 2SW, UK

Tel: 0455 212055

FTW Magnetos & Dynamos
6 Winster Road, Sheffield, Yorkshire (S.), S6, UK

Tel: 0742 336267

Magneto and Dynamo overhauls, alternator rewinds, K-Tec solid state regulators. All work guaranteed two years free estimates. Opening hours 10.30am to 5.50pm, Monday to Friday 11.00am to 1.00pm Saturday.

Fast Friends
PO Box 19, Fishponds, Bristol, Avon, UK

Federation of European Motorcyclists
45 St Katherines Road, Whipton, Exeter, Devonshire, EX4 7JW, UK

Fellowship of Historic Motorcycle Assocs
75 Boulton Grange, Randlay, Telford, Shropshire, TF3 2LD, UK

Contact: John Law

Frank Thomas Limited
Midland Road, Higham Ferrers, Northamptonshire, NN9 8DN, UK

Tel: 0933 410272

Frank Webber
96 Bellhouse Road, Eastwood, Leigh on Sea, Essex, SS9 5NG, UK

Tel: 0702 511067

Fred Barlow
Unit 5, Beauchamp Ind.Park, Watling Street, Two Gates, Tamworth, Staffordshire, UK

Tel: 0827 251112

Front Line Motorsports
48 Middlesbrough Road, Middlesbrough, Cleveland, UK

Tel: 0642 452426

G & D Plating & Polishing
Clothalbury Farm Ind.Estate, Baldock, Hertfordshire, SG7 6RJ, UK

Tel: 0462 79729

Gatefield Classic Motorcycles
Unit 2, Wellington Road, Ashton under Lyme, Lancashire, UK

Tel: 061 343 4774

H E Webber & Sons
91 Kenton Road, Kenton, Harrow, Middlesex, HA3 0AN, UK

Tel: 081 907 0041

Halco Tuning
Easter Works, Sutton Mandeville, Salisbury, Wiltshire, SP3 5NL, UK

Tel: 0722 714700

The Yamaha XS650 specialists. Everything from spares to tuning and race parts. Engine rebuilds and exhaust systems. Bikes from "projects" to "mint" standard examples.

Hewy Design Paintwork
Spong Hill Farm, North Elmham, Dereham, Norfolk, NR20 4DF, UK

Tel: 0362 668068

Quality airbrush design helmets a speciality, repairs, resprays, and plastic welding. Mon-Sat 9-6. Postal service available.

Hobbsport Racing
Unit 6 Morderna Business Park, Mytholmroyd, Hebden Bridge, Yorkshire (W.), HX7 5QQ, UK

Tel: (0422) 884978

House of Kolor
54 Denmark Road, Winton,
Bournemouth, Dorset,
BH9 1PB, UK

Tel: 0202 526477

Hughenden M40
Milton Common,
Oxfordshire, UK

Contact: Chris Hooper
Tel: 0844 279701

Hughie Hancox
21 Bayton Road, Exhall,
Coventry, Midlands (W.),
UK

Tel: 0203 368038

Hydrocontrol Systems
Chapel Works,
Laddingford, Maidstone,
Kent, ME18 6BT, UK

Tel: 0622 872146

Impact Signs
Atlantic Business Centre,
Altrincham, Cheshire,
WA14 5NQ, UK

Tel: 061 929 9594

IWF Ltd
78A Forsyth Road,
Newcastle upon Tyne, Tyne
& Wear, NE2 3EU, UK

Tel: 091 281 0945

J Morgan
Brookfield Mill, Crumlin
Road, Belfast, Co. Antrim,
BT14 7EA, UK

Tel: 0232 757720

Jack Bottomley's
John Street, Garden Lane,
Salford, Manchester, Gt.,
M3 7AU, UK

Tel: 061 834 7799

Joe Francis Motors Ltd
340 Footscray Road, New
Eltham, London, SE9 2ED,
UK

Tel: 081 850 1373
Fax: 081 859 5617

John Allaway & Sons
43 The Street,
Wrecclesham, Farnham,
Surrey, UK

Tel: 0252 725335

John Hall and Sons
102-108 Devonshire Road,
Blackpool, Lancashire, UK

Tel: 0253 393320

John Harris Motorcycles
Whitehall Road,
Crowborough, Sussex (E.),
UK

Tel: 0892 652380

Johns of Romford
46-52 London Road,
Romford, Essex, RM17
9QX, UK

Tel: 0708 746293

KAS Metal Polishing
Unit 8, New Road Business
Estate, Ditton, Aylesford,
Kent, ME20 6AF, UK

Tel: 0732 871186

Keith Dixon Motorcycles
392-396 Blackburn Road,
Accrington, Lancashire,
BB5 1SA, UK

Tel: 0254 231221

Kirby Rowbotham
58 Arch Street, Rugeley,
Staffordshire, WS15 1DL,
UK

Tel: 0889 584758

Lance Rayner
16 Cambridge Road,
Barton, Cambridge,
Cambridgeshire, CB3 7AR,
UK

Tel: 0223 262888

Leo Yuill
"Amancay", Crows Green,
Bardfield Sailing, Essex,
CM7 5EB, UK

Tel: 0371 850477

Lewis & Templeton
Brinklow Service Station, 26
Coventry Road, Brinklow,
Rugby, Warwickshire,
CV23 ONE, UK

Tel: 0788 833330

Lightning Bolt Co.
P.O.Box 69, Rochester,
Kent, ME1 1XX, UK

Contact: Chris Bloomfield
Tel: 0634 271276
Fax: 0634 271276

The stainless steel fasteners
(nuts, bolts etc) specialist for
Harleys. Screwkits and
individual UNF and UNC
fasteners. Specials e.g headbolts.
Free price list and advice.

Lightning Spares
157 Cross Street, Sale,
Cheshire, M33 1JW, UK

Tel: 061 969 3850

Long Eaton Enamellers
Acton Avenue, Long Eaton,
Nottinghamshire, UK

Tel: 0602 723676

MFS Polishing & Plating
Unit 3 & 4 Park Farm Ind
Est, Wood Lane, Ramsey,
Cambridgeshire, PE17 1XF,
UK

Tel: 0487 711103
Fax: 0487 815250

MH Racing Services
Unit C1, Fiveways Ind.
Estate, Rudloe, Corsham,
Wiltshire, SN13 ONX, UK

Tel: 0225 811583

Magnum
70 Worcester Street,
Wolverhampton, Midlands
(W.), WV2 4LE, UK

Tel: 0902 27915

Military Vehicle Trust
PO Box 6, Fleet,
Hampshire, GU13 9PE, UK

**Motad International
Limited**
P.O.Box 84, Hospital
Street, Walsall, Midlands
(W.), WS2 8TR, UK

Tel: 0922 725 59
Fax: 0922 640379

Moto Imperial Supplies
Unit 3, City Estate,
Corngreaves Road, Cradley
Heath, Warley, Midlands
(W.), B64 7EP, UK

Tel: 0384 411482
Fax: 0384 411919

**Motorcycle Association
of GB**
Starley House, Eaton Road,
Coventry, Midlands (W.),
UK

**Motorcycle Retailers
Association**
201 Great Portland Street,
London, W1N 6AB, UK

Tel: 071 580 9122

NCK Racing
339 Bedworth Road,
Longford, Coventry,
Midlands (W.), CV6 6BN,
UK

Tel: 0203 362334

NLVR
P.O.Box 1455, Windsor,
Berkshire, SL4 1QR, UK

Tel: 0753 831553

**National Motorcycle
Council**
Cowley House, 9 Little
College Street, London,
SW1E 3XS, UK

*Tel: 071 222 0664
Fax: 071 233 0335*

**National Scooter Riders
Association**
34 Coronation Road,
Forest Town, Mansfield,
Nottinghamshire,
NG19 0AJ, UK

Contact: Jeff Smith

P&J Powder Coatings
17 Evanton Place,
Thornliebank Ind. Estate,
Glasgow, Strathclyde, G46,
UK

Tel: 041 620 1652

**PHD Metal
Refurbishment**
Unit 7, Delta Way
Bus.Centre, Longford Road,
Bridgeton, Cannock,
Staffordshire, UK

Tel: 0543 462590

Paul Savage
Unit 17,Enterprise Works,
28 Hemming Rd, Washford
Est., Redditch, Hereford &
Worcester, B98 0DH, UK

Tel: 0527 21666

Penrite Oil
Unit 1B, 31 Dollman Street,
Birmingham, Midlands (W.),
B7 4RP, UK

Peter Bond
31 Avington Grove,
London, SE20 8RY, UK

Tel: 081 659 1396

Peter Watmough
40 Maylea Drive, Otley,
Yorkshire (W.), LS21 3ND,
UK

Tel: 0943 850239

Phil's British Bike Bits
204 Wellingborough Road,
Rushden,
Northamptonshire, UK

Tel: 0933 53243

Phillips Transfers Ltd
15 Stock Road, Billericay,
Essex, UK

Tel: 0277 659505

R A V Engineering
2 Gorsey Lane, Banks,
Southport, Merseyside,
PR8 8EH, UK

Tel: 0704 212826

RCM
Unit 15, Argyle Way, Ely,
Cardiff, Glamorgan (S.), UK

Tel: 0222 593122

R. D. Cox & Son
Phoenix Works, Phoenix
Terrace, Hartley Wintney,
Hampshire, RG27 8RU, UK

Tel: 0252 845352

RAC Motoring Services
PO Box 700, Spectrum,
Bond Street, Bristol, Avon,
BS99 1RB, UK

Racespec
Yauncos Cottage, Tillers
Green, Dymock,
Gloucestershire, GL18 2AP,
UK

Tel: 0531 890250

Renntec
69 Woolsbridge Ind.Estate,
Three Legged Cross,
Wimbourne, Dorset,
BH21 6SP, UK

Tel: 0202-826722

Rick James
1a Suggits Lane,
Cleethorpes, Humberside
(S.), DN35 7JE, UK

Tel: 0472 694468

Ride On Motorcycles Ltd
19-21 Nithsdale Street,
Glasgow, Strathclyde, UK

Tel: 041 424 0404

Robin Watson Signs
23a Princes Street,
Corbridge,
Northumberland,
NE45 5DE, UK

Tel: 0434 632089

Robspeed Motorcycles
26 Cross Street,
Cleethorpes, Humberside,
UK

Tel: 0472 602208

Rock Oil Company
PO Box 155, Wharf Street,
Warrington, Cheshire,
WA7 7HU, UK

*Contact: Charles Hewitt
Tel: 0925 36191
Fax: 0925 32499*

Rock Oil are an independent
company supplying high quality
lubricants to the motorcycle
industry & currently exporting
worldwide.

Ron Jenkins
24 Stanwell Drive,
Middleton Cheney,
Northamptonshire,
OX17 2RB, UK

Tel: 0295 710740

Rugeley Motorcycles
58/59 Horsefair, Rugeley,
Staffordshire, WS15 2EJ,
UK

Tel: 0889 585085

SAMS
PO Box 6, Withernsea,
Yorkshire (E.), HU19 2HJ,
UK

Tel: 0964 612235

SB Products
29 Wulfric Square, Bretton,
Peterborough,
Cambridgeshire, PE3 4RF,
UK

Tel: 0733 266138

STW Motorcycles
178 Penistone Road North,
Sheffield, Yorkshire (S.),
S6 1QA, UK

Tel: 0742 336269

Scott Oilers
106 Clober Road,
Milngavie, Glasgow,
Strathclyde, G62 7SS, UK

Tel: 041 956 6225

Screencraft Ltd
17 Lombard Road,
Wimbledon, London, UK

Tel: 081 543 8977

Silks Solicitors
Barclays Bank Chambers,
Birmingham Street,
Oldbury, Midlands (W.),
B69 4EZ, UK

*Tel: 021 511 2233
Fax: 021 552 6322*

Silks solicitors - for legal advice
on personal injury folowing an
accident, please contact Martin
Bradley.

Skidmarx
Unit 2, Waggoners Yard,
Baldock Street, Ware,
Hertfordshire, UK

Tel: 0920 487547

Sports & Vintage Motors
Upper Battlefield,
Shrewsbury, Shropshire,
SY4 3DB, UK

Tel: 0939 210458

Stanford Auto Factors
Unit 1, Uffington Trading Estate, Uffington, Oxfordshire, SN7 7QD, UK

Tel: 0367 7106187

Stewart Engineering
Church Terrace, Harbury, Leamington Spa, Warwickshire, CV33 9HL, UK

Tel: 0926 612589

Suppliers and manufacturers since 1960 of S7 and S8 Sunbeam motorcycle spares. Complete engineering and restoration facilities all available in house.

Sunami Motorcycles
14-15 Morden Court, Morden, Surrey, UK

Tel: 081 646 1554

Surrey Cycles
Surrey House, High Street, Cranleigh, Surrey, GU6 8RL, UK

Tel: 0483 272328

Amal-Mki Concentric carburettors and spares.

Swann & Moore (Assessors) Ltd
83a St John's Way, Corringham, Stanford-le-Hope, Essex, SS17 7LL, UK

Tel: 0375 640166
Fax: 0375 644098

Accident Claims Consultants operating on "No Win, No Fee" basis. 30 years experience handling claims for compensation arising from road accidents. Motorcyclists a speciality.

TSP Services
178-180 Dukes Ride, Crowthorne, Berkshire, RG11 6DS, UK

Tel: 0344 761662/762291

Any motorcycle supplied nationwide - hundreds of used all around the country. Find yours the easy way. Monday to Friday 9am to 6pm.

Taylor Racing
23-25 Station Hill, Chippenham, Wiltshire, UK

Tel: 0249 657575/6

Technique Tyres Ltd
4 Kelvin Road, Thundersley, Essex, UK

Tel: 0268 795522

The Finishing Touch
26 East Hanningfield Rd, Rettendon, Chelmsford, Essex, CM3 8EQ, UK

Tel: 0243 400918

Classic motor cycle paintwork, to restore rust, dents, pitting. Tank repairs, chroming, colour matching cellulose - two pack. Paint lining - wheels painted and lined. Complete motorcycles or parts restored.

The Goodman Engineering Co. Ltd
Westward, Buckle St, Honeybourne, Evesham, Hereford & Worcester, WR11 5QQ, UK

Tel: 0386 832090
Fax: 0386 831614

The Paint Studio
95 Station Road, Ilkeston, Derbyshire, DE7 5LL, UK

Tel: 0602 322290

Tony Hayward
28 Kelsterton Road, Connah's Quay, Deeside, Clwyd, CH5 4BJ, UK

Tel: 0244 830776

Trail Riders Fellowship
18 Corsham Road, Fords Farm, Calcott, Reading, Berkshire, RG3 5ZH, UK

Contact: *Keith Forthergill*

Tran Am Ltd
William House, Gore Road, New Milton, Hampshire, BH25 6RJ, UK

Tel: 0425 620580

Triskill Ltd
14 Bissell Street, Birmingham, Midlands (W.), B5 7HP, UK

Tel: 021 622 3294
Fax: 021 622 5085

Triumph Designs Limited
Jacknell Road, Dodwells Bridge Ind. Estate, Hinckley, Leicestershire, LE10 3BS, UK

Tel: 0455 251600
Fax: 0455 251367

UK Motorcycle Exports
16 Begonia Drive, Burbage, Leicestershire, LE10 2SW, UK

Tel: 0455 611030

WDB Engineering Service
75 High Street, Winslow, Buckinghamshire, MK18 3DG, UK

Tel: 0296 712906

Walker Engineering
P.O.Box 100, Halifax, Yorkshire (W.), UK

Tel: 0422 345568

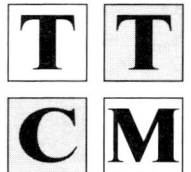

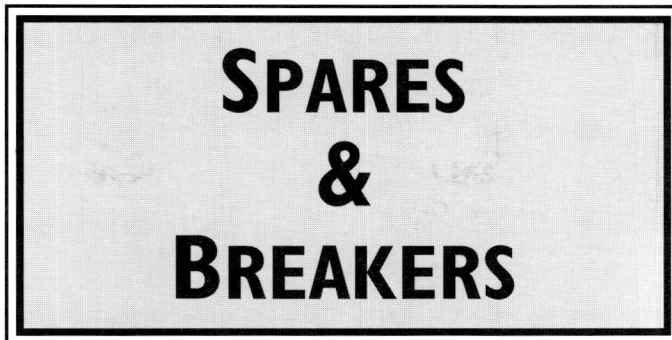

SPARES & BREAKERS

A1 Bikes
Unit 2K, Godington Way
Ind Est, Ashford, Kent, UK

Tel: 0233 638648

Abbey Motorcycles
28 Bradwall Road,
Sandbach, Cheshire,
CW11 9AJ, UK

Tel: 0270 760138

Alpha
Kingsley Street, Netherton,
Dudley, Midlands (W.), UK

Tel: 0384 255151

Amal Concentric Carburettors
70 Blakemore Drive, Sutton
Coldfield, Midlands (W.),
B75 7RW, UK

Tel: 021 378 0241

Archway Motorcycles
18 Blandford Square,
Newcastle upon Tyne, Tyne
& Wear, UK

Tel: 091 232 9284

Ariel
The Old Town Maltings,
Broad Street, Bungay,
Suffolk, NR35 1EE, UK

Tel: 0986 894798

Armours
784 Wimborne Road,
Bournemouth, Dorset,
BH9 2HS, UK

Tel: 0202 519409
Fax: 0202 5510671

Classic exhaust, silencer
manufacturer, stainless, chrome,
British, Continental, Classic
tyres, mudguards, and general
cycle parts. Export welcome.
Closed Mondays. Catalogue
£1.00.

Armstrong Engineering
176 Westgate Road,
Newcastle upon Tyne, Tyne
& Wear, NE4 6AL, UK

Tel: 091 261 4579

B & C Express Products
Station Road,
Potterhanworth, Lincoln,
Lincolnshire, LN4 2DX, UK

Tel: 0522 791369

Backpackers Club
20 St.Michael Road,
Tilehurst, Reading,
Berkshire, RG3 4RP, UK

Bradford Ignition Services
Unit 6, Thorncliffe Ind.Est.,
Thorncliffe Road, Bradford,
Yorkshire (W.), BD8 7DD,
UK

Tel: 0274 546268

Bartel Alloy Tanks
Unit 8, 295 Killaughey
Road, Donaghadee,
Co.Down, BT21 0LY, UK

Tel: 0247 820786

Benelli (Brembo Club International)
21 Mount Pleasant, Sutton
in Ashfield,
Nottinghamshire, UK

Contact: P Rimmer

Bike Busters
Wootton Rivers,
Marlborough, Wiltshire, UK

Tel: 0672 810319

Bike Spurz
27 Junction Lane, Sutton,
St Helens, Merseyside,
WA9 3JN, UK

Bike-Teck
Limes Garages, A22
Eastbourne Road, Blindley
Heath, Surrey, RH7 6JJ, UK

Bill Head Ltd
Southgate, North Road,
Preston, Lancashire,
PR1 1NP, UK

Tel: 0772 52066
Fax: 0772 562160

Blaise Motorcycles Ltd
242 Stapleton Road,
Eastville, Bristol, Avon,
BS5 0NT, UK

Tel: 0272 355184

Blays
32-38 The Green,
Twickenham, Middlesex,
UK

Tel: 081 894 2103

Bob Jackson
Rose & Crown Works,
Stricklandgate, Kendal,
Cumbria, UK

Tel: 0539 720582

Bob Porecha
303 Sydenham Road,
Sydenham, London,
SE26 5EW, UK

Tel: 081 659 8860
Fax: 081 659 9196

Bob Price Classic Bikes
14 The Crescent, Leek,
Staffordshire, ST13 6HD,
UK

BSA A50/65 unit twin specialist.
New and used parts.
Restorations, engine rebuilds.
Write or phone Bob on 0538
385939 daytime 1pm-7pm.

Brian Bennett
13 Derby Road Business
Park, Derby Road, Burton
on Trent, Staffordshire,
DE14 1RW, UK

Tel: 0283 511841

Bristol Breakers
Castle House, Ducie Road,
Lawrence Hill, Bristol,
Avon, BS5 0AT, UK

Tel: 0272 351133

Britbits
185 Barrack Road,
Christchurch, Dorset,
BH23 2AT, UK

Tel: 0202 483675
Fax: 0202 475327

New restoration spares for post
war Triumph, BSA and Norton
motorcycles. Established 35
years. Closed Mondays and
Tuesday (mail order only).

British Marketing
27324 Camino Capistrano
No.139, Laguna Niguel,
California, 92677, USA

Tel: 714 582 2902

British Motorcycle Engineering
A4 Penarth Dock, Penarth,
Glamorgan (S.), CF6 1LR,
UK

Tel: 0222 708132

Burton Bike Bits
138 Waterloo Street,
Burton on Trent,
Staffordshire, DE14 2NF,
UK

Tel: 0283 34130
Fax: 0283 511096

Specialists in British classic
motorcycles and spares. Also
repairs and restorations. Royal
Enfield, BSA, Triumph, Norton,
Enfield and India. Open
Mon-Sat, 9-6.

C J Wilson
23-25 West Main Street,
Uphall, Broxburn, Lothian,
UK

Tel: 0506 856751

CTG Racing Ltd
8 Telford Road, Ferndown
Ind. Estate, Wimborne,
Dorset, BH21 7QL, UK

Tel: 0202 871102

Camelford Bike Bits
35 High Street, Camelford,
Cornwall, UK

Tel: 0840 213483

Carl Rosner Ltd
Station Approach,
Sanderstead Road, South
Croydon, Surrey, CR2
OPL, UK

Contact: Carl Rosner
Tel: 081 657 0121
Fax: 081 651 0596

Established 25 years.
Comprehensive spares stock for
Commando, Triumph 650/750.
Trident/R-3. Full workshop
facilities, including restorations.
Dealers for Hinckley Triumphs,
Norton Rotary and Indian
Enfield.

**Central Wheel
Components Ltd**
Lichfield Road, Water
Orton, Birmingham,
Midlands (W.), B46 1NU,
UK

Tel: 021 749 5511

**Chadwell Motorcycles of
Essex**
101 Riverview, Chadwell St
Mary, Grays, Essex,
RM16 4BD, UK

Tel: 0375 842601

Charlie's
169-171 Fishponds Road,
Eastville, Bristol, Avon,
BS5 6PR, UK

Tel: 0272 511019

**Charnwood Classic
Restorations**
107 Central Road,
Hugglescote, Coalville,
Leicestershire, LE5 2FL, UK

Tel: 0530 832257

Chris Applebee
471 Rayleigh Road,
Benfleet, Essex, UK

Chris Elderfield
25 Church Street,
Goldalming, Surrey, GU7
1EL, UK

Tel: 0483 427940

**Classic & Custom
Motorcycles**
125 Chester Road,
Northwich, Cheshire,
CW8 4AA, UK

Tel: 0606 77262
Fax: 0606 77262

Pitted forks? Forks ground and
hardchromed to the highest
standards. Nine day turnround;
Securicor collection and delivery
guaranteed better than new!
Faxed information sheet
available. Callers by prior
arrangement only.

**Classic & Custom
Motorcycles**
(Harley-Davidson spares -
ring/fax only)

Contact: Dave
Tel: 0860 312530
Fax: 0606 77262

Harley-Davidson spares. We buy
and sell second hand Harley-
Davidson spares from 1940's to
present. See our stall at
autojumbles around the country.

**Clubman Racing
Accessories**
P.O.Box 59, Fairfield,
Connecticutt, CT06430,
USA

Tel: 203 256 1224

Conway Motors
224 Tankerton Road,
Whitstable, Kent,
CT5 2AY, UK

Tel: 0227 276405
Fax: 0227 771363

Crankshaft Specialists
39 Sideley Road, Kegworth,
Derbyshire, DE74 2FJ, UK

Tel: 0509 673295

D&K
Riverside Works, Station
Road, Cheddleton,
Staffordshire, ST13 7EE, UK

Tel: 0538 361117

DMW Motor Cycles
Valley Road Works,
Sedgley, Dudley, Midlands
(W.), UK

Tel: 880351

D. Wilkinson
49 Mansfield Road,
Warsop, Mansfield,
Nottinghamshire,
NG20 OEQ, UK

Tel: 0623 843963

Dave Lindsley
196 Pilsworth Road,
Heywood, Lancashire,
OL10 3DY, UK

Tel: 0706 365838
Fax: 0706 627500

Dave Mackenzie
213/217 Haughton Road,
Darlington, Co.Durham, UK

Tel: 0325 465045

David Payne M/Cs
Irvine House, Northgate,
Whiteland Industrial Estate,
Morecambe, Lancashire, UK

Tel: 0524 846254

Daytona
42-48 Windmill Hill, Ruislip
Manor, Middlesex,
HA4 8PT, UK

Tel: 0895 675511
Fax: 0895 630654

Dealership for Kawasaki,
Triumph and Ducati
motorcycles. Also main parts
stockist for Kawasaki.

Disc Pads
Elberton Garage, Camp
Lane, (B4461),Elberton,
Bristol, Avon, BS12 3AQ,
UK

Tel: 0454 418676

E. Jenkins
90 Elliman Avenue, Slough,
Berkshire, SL2 5BE, UK

Tel: 0753 529313

Eddie Two Wheels
168 Cross Street, Sale,
Cheshire, UK

Tel: 061 973 8868

Exe Bike Breakers
Unit 1, 6 Marsh Green
Road, Exeter, Devonshire,
EX2 8NY, UK

Tel: 0392 413451

**Export Sales & Mail
Spares**
The Firs, Othery,
Somerset, UK

Tel: 0823 698305

Express Bike Spares
69 May Lane, Hollywood,
Birmingham, Midlands (W.),
B47 5PA, UK

Faithfull Classics
76 Parkway, Wickham
Market, Woodbridge,
Suffolk, IP13 OSS, UK

Tel: 0728 747763

Vintage & classic motorcycle
spares. Mail order.

Ferodo
1 High Street, Earl Shilton,
Leicestershire, UK

Tel: 0455 841133

**Ferodo Classic Brake
Services**
PO Box 5, Whaley Bridge,
Stockport, Cheshire, SK12
7LL, UK

Tel: 0663 732940

Francis Motors
173/A Burton Road,
Lincoln, Lincolnshire,
LN1 3LW, UK

Tel: 0522 537324

**Fred Cheshire
Motorcycles Ltd**
19-23 Prestbury Road,
Cheltenham,
Gloucestershire,
GL52 2PN, UK

GWD M/C Spares
63 Main Street, Leuchars,
Fife, UK

Tel: 033 483 8289
Fax: 033 483 8289

Geoff Dodkin
346 Upper Richmond Road
West, East Sheen, London,
SW14 7JS, UK

Tel: 081 876 8779

George Pew
Millhouse, Barkway,
Royston, Hertfordshire,
SG8 8BX, UK

Tel: 0763 848763

**Gibson Motorcycle
Breakers**
9-13 Gladstone Lane,
Scarborough, Yorkshire
(N.), YO12 7BP, UK

Tel: 0723 500041

Gowrings
245 Finchampstead Road,
Wokingham, Berkshire, UK

Tel: 0734 770350

Grange Works
Burnley Road, Rawtenstall,
Rossendale, Lancashire,
BB4 8HY, UK

Tel: 0706 226127

Grimeca
462 Station Road,
Dorridge, Solihull,
Midlands (W.), UK

Tel: 056477 5835
Fax: 0564 770461

Hagon Products Ltd
350 High Road, Leyton,
London, E10 6QQ, UK

Tel: 081 556 4447

Manufacturers of Hagon
suspension units, direct supply
of replacement units for nearly
all motorcycles; also wheel rims,
spokes and wheelbuilding. Cast
wheel repairs.

Hamrax Motors Ltd
328 Ladbroke Grove,
North Kensington, London,
W10 5AD, UK

Tel: 081 969 5380
Fax: 081 960 1993

Harlgo Limited
462 Station Road,
Dorridge, Solihull,
Midlands (W.), UK

Tel: 056477 5835

**Harwoods of Richmond
Ltd**
14-18 Kew Foot Road,
Richmond, Surrey,
TW9 2SS, UK
Tel: 081 940 2045

Hemming & Wood Ltd
Unit 9, Shires Industrial Est,
Birmingham Road, Lichfield,
Staffordshire, WS14 9BW,
UK

Tel: 0543 256711

Hitchcock's Motorcycles
Long Close, Glasshouse
Lane, Hockley Heath, Mid-
lands (W.), B94 6PZ, UK

Tel: 0564 783192
Fax: 0564 783313

Hobbs Sport Earls Ltd
4D Brent Mill Ind.Estate,
South Brent, Devonshire,
TQ10 9YT, UK

Tel: 0364 73956
Fax: 0364 73957

Hobbsport Racing
Unit 6 Morderna Business
Park, Mytholmroyd,
Hebden Bridge, Yorkshire
(W.), HX7 5QQ, UK

Tel: (0422) 884978

**Hog & Chop
International**
48 Manchester Street,
Cleethorpes, Humberside
(S.), DN35 7QG, UK

Tel: 0472 697859

Hunters
241 Westgate Road,
Newcastle upon Tyne, Tyne
& Wear, NE4 6AE, UK

Tel: 091 261 8592

Husky Sport
Cagiva Sport, 35 Long Shot
Lane, Bracknell, Berkshire,
RG12 1RL, UK

Tel: 0344 56860

Impact Signs
Dept SB, Atlantic Business
Centre, Altrincham,
Cheshire, WA14 5NQ, UK

Tel: 061 929 9594

J.W. Tennant-Eyles
Barcote Manor, Buckland,
Nr.Faringdon, Oxfordshire,
SN7 8PP, UK

Tel: 036 787 330

JB's
Unit 18, Rake Ind.Estate,
Canhouse Lane, Rake,
Petersfield, Hampshire, UK

Tel: 0730 895550

James Wheildon
Old Brickwood Farm, West
Grimstead, Salisbury,
Wiltshire, SP5 3RN, UK

Tel: 0722 72701

Jim Matthews
The Old Chapel, Sun Lane,
Crich, Derbyshire,
DE4 5BR, UK

Tel: 0773 856657

Joe Shaw & Son
6 Marion Close, Chatham,
Kent, ME5 9QA, UK

Tel: 0634 861552/862793
Fax: 0634 868308

John Timson Motorcycles
Unit B, Electron Works,
275 Brook Street, Preston,
Lancashire, PR1 7NH, UK

Tel: 0772 204633

Johns of Romford
46-52 London Road,
Romford, Essex,
RM17 9QX, UK

Tel: 0708 746293

Julian Soper Motorcycles
1335 London Road, Leigh
on Sea, Essex, SS9 2AB, UK

Tel: 0702 715727
Fax: 0702 471360

Specialists in new spares and
accessories for all Japanese bikes,
50cc to 1500cc, 1970 to 1993.

K & J Gardner
Wolford Heath, Shipston
on Stour, Warwickshire,
CV36 5RN, UK

Tel: 0608 84 306
Fax: 0608 84 785

KAIS
Punchbowl Garage,
Atherton, Manchester,
Manchester, Gt., UK

Tel: 0942 896366

**Kidderminster
Motorcycles**
60/61 Blackwell Street,
Kidderminster, Hereford &
Worcester, UK

Tel: 0562 66679
Fax: 0562 825826

KTM Motorcycles
10 High Street, Sileby,
Leicestershire, UK

Tel: 0509 816177

Kawasaki Spares
Courier Road, Phoenix
Parkway, Corby,
Northamptonshire,
NN17 1DY, UK

Keith Benton
Via Gellia, Kirkhead Road,
Grange-Over-Sands,
Cumbria, LA11 7DD, UK

Keith Dixon Motorcycles
392-396 Blackburn Road,
Accrington, Lancashire,
BB5 1SA, UK

Tel: 0254 231221

Ken Jackson
19 High Street,
Whitchurch, Aylesbury,
Buckinghamshire,
HP22 4JU, UK

Tel: 0296 641815
Fax: 0844 216269

Lambrook Tyres Ltd
Farway, Colyton,
Devonshire, EX13 6DL, UK

Tel: 0404 87282
Fax: 0404 87477

Lambrook tyres is one of the
UK's major suppliers of tyres for
all old motorcycles. Many rare
and unusual sizes in stock.

Lightning Bolt Co.
P.O.Box 69, Rochester,
Kent, ME1 1XX, UK

Contact: Chris Bloomfield
Tel: 0634 271276
Fax: 0634 271276

The stainless steel fasteners
(nuts, bolts etc) specialist for
Harleys. Screwkits and
individual UNF and UNC
fasteners. Specials e.g headbolts.
Free price list and advice.

Lightning Spares
157 Cross Street, Sale,
Cheshire, M33 1JW, UK

Tel: 061 969 3850

London Suzuki Centre
859-861 Harrow Road,
College Park, London,
NW10 5NH, UK

Tel: 081 969 0741

**M R Holland
(Distributors) Ltd**
Unit 2, Benner Road,
Wardentree Lane Ind Est,
Spalding, Lincolnshire,
PE11 3UG, UK

Tel: 0775 766455
Fax: 0775 710292

Shock absorbers and fork springs
for most makes available ex
stock. Mail order and trade
welcome. Specialists in BMWs,
Harleys, Goldwings.

M&P Accessories Ltd
Unit 3, Wards Farm Ind.Est.,
Woodcote, Reading,
Berkshire, UK

Tel: 0491 681487

M. R. D. Metisse
London Street, Kingswood,
Bristol, Avon, BS15 1QZ,
UK

Tel: 0272 600893

MPS
Daneheath Business Park,
Heathfield, Newton Abbot,
Devonshire, TQ12 6TL, UK

Tel: 0626 835835
Fax: 0626 835152

The UK's largest mail order motorcycle house, products stocked
include D.I.D. chain, Vesram
disc pads, Motad exhausts,
Goodridge brake lines, Moto
Fizz accessories, Abus locks.

MXB Motocross
Wonastow West Ind.Estate,
Monmouth, Gwent,
NP5 3AH, UK

Tel: 0600 772211

Maico Spares
16 Nunts Park Avenue,
Coventry, Midlands (W.),
CV6 4GY, UK

Tel: 0203 366317

Manx Spare Parts
P.O.Box 12-500, Penrose,
Auckland, New Zealand

Tel: 64 9 570 1119

Martyn's Motorcycles
5 Hotel Street, Coalville,
Leicestershire, LE6 2EQ, UK

Tel: 0530 810308

McIntosh Racing
P.O.Box 12-500, Penrose,
Auckland, New Zealand

Meeten and Ward Ltd
360 Kingston Road, Ewell,
Surrey, KT19 0DU, UK

Tel: 081 393 5193
Fax: 071 947 2595

Mettbikes Breakers
Unit 5, Ivory Estate, West
Hendon Broadway, London,
NW9, UK

Tel: 081 202 9260

Michael Freeman Motors
The Camp, Stroud, Glou-
cestershire, GL6 7HN, UK

Tel: 0285 821297

Mick Hemmings
72/74 Overstone Rd,
Northampton, Northamp-
tonshire, NN1 3JS, UK

Tel: 0604 38505
Fax: 0604 31838

Mike's Bike Spares
Unit 3A, Plot F,
Westminster Est., Station
Rd., North Hykeham,
Lincolnshire, LN6 3QY, UK

Tel: 0522 689905

**Motad International
Limited**
PO Box 84, Hospital Street,
Walsall, Midlands (W.),
WS2 8TR, UK

Tel: 0922 725559

Moto Cinelli
Tweed Road, Weedon
Rd.Ind.Est, Northampton,
Northamptonshire, UK

Tel: 0604 750851

Moto-Bins
16 Surfleet Road,, Surfleet,
Spalding, Lincolnshire,
PE11 4AQ, UK

Tel: 0775 85881

Motorworks
Thirston Works, Thirston
Road, Honley,
Huddersfield, Yorkshire
(W.), HD7 2JH, UK

Tel: 0484 664511
Fax: 0484 667229

BMW parts, new and used. Ring
for free catalogue. Open 9.00 to
5.30, Mon-Fri.

Motorworks
Units 1-3 Tanyard
Ind.Estate, Milnsbridge,
Huddersfield, Yorkshire
(W.), HD3 4NB, UK

Tel: 0484 640030
Fax: 0484 460383

Motrac Racing Spares
18 Westminster Ind.Park,
Ellesmere Port, South
Wirral, Cheshire,
L65 3DU, UK

Tel: 051 357 1062

Mr Fastner Ltd
Unit 2, Warwick House
Industrial Park, Banbury
Road, Southam, Warwick-
shire, CV33 0HL, UK

Tel: 0926 817207

NDT Motorcycles
Banks Bldgs., Front Street,
New Herrington, Houghton
le Spring, Tyne & Wear,
DH4 7AU, UK

Tel: 091 584 8881

Neil Hudson Motorcycles
8 Old Tarnwell, Station
Drew, Bristol, Avon, BS18
4EA, UK

Tel: 0275 333244

Neil Young Motorcycles
Barrack Road,
Northampton,
Northamptonshire, UK

Tel: 0604 22411

Norman White Norton
Thruxton Circuit, Andover,
Hampshire, UK

Tel: 0264 773326

Nova
8 Horseshoe Yard,
Crowland, Linconshire,
PE6 OBJ, UK

Tel: 0733 210082

P Kemble
Ruther Glen, Birdham,
Chichester, Sussex (W.),
UK

Tel: 0243 511448

Pedal & Motor Limited
P.O.Box 190, Southampton,
Hampshire, SO9 7YD, UK

Peter Jones
7 Frankwell, Shrewsbury,
Shropshire, SY3 8JY, UK

Tel: 0743 357375

**Philip Youles
Motorcycles**
Belgarth Road Works,
Accrington, Lancashire,
BB5 6AH, UK

Tel: 0254 234051
Fax: 0254 386617

Large selection of quality used
spares supplied off the shelf,
MOTs, rebores, pattern spares
and general servicing work
carried out on all makes, also
MZ main agents with
four-strokes in stock. Open
9am-7pm, Mon-Fri, 9am-5pm Sat.

Phoenix
High Woodland, Littlegain,
Hilton, Bridgend,
Shropshire, WV15 5PA, UK

Tel: 0384 221440

Powerbronze
44 Brook Lane, Ferring,
Worthing, Sussex (W.), UK

Tel: 0903 507300

Pratts
166 Harcourt Avenue,
Sidcup, Kent, DA15 9LW,
UK

Tel: 081 302 2133

Premier Plates
43 West Street, Riddings,
Derbyshire, UK

Tel: 0773 747295

RJ Motorcycles
18-20 Hotel Street,
Coalville, Leicestershire,
UK

Tel: 0530 833297

RIP Spares
38-40 Lawn Lane, Hemel
Hempstead, Hertfordshire,
UK

Tel: 0442 234715

Racespec
Yauncos Cottage, Tillers
Green, Dymock,
Gloucestershire, GL18 2AP,
UK

Tel: 0531 890250

Rafferty Newman Ltd.
32 Exmouth Road,
Southsea, Hampshire,
PO5 2QL, UK

Tel: 0705 755125

Raven Cycles
Brooklands, Masons Bridge
Road, Redhill, Surrey, UK

Tel: 0737 762361

Ray Dentith
6 London Road, Purbrook,
Waterlooville, Hampshire,
UK

Tel: 0705 230746

Redcar Motorcycles
2 Redcar Road, Belgrave,
Leicester, Leicestershire,
LE4 6PD, UK

Tel: 0533 661637

Reg Allen
37/41 Grosvenor Rd,
Hanwell, London, W7, UK

Tel: 081 567 1974
Fax: 081 579 1248

London's largest stockist of
Meriden Triumph spares;
machines bought, sold and
exchanged.

**Rob Willsher Suzuki
Spares**
Victoria Garage (Burlesdon),
Southampton, Hampshire,
SO3 8ES, UK

Tel: 0703 404730
Fax: 0703 405087

Suzuki parts specialists. Phone
for mail order. Visa and access
welcomed.

Roebuck Motorcycles
354 Rayners Lane, Pinner,
Middlesex, HA5 5EN, UK

Tel: 081 868 1231

Rogersons
Orrell Post, Wigan,
Lancashire, UK

Tel: 0942 212221

Ron Kemp
Ty Vin, Llanddewi,
Llandrindod Wells, Powys,
LD1 6SE, UK

Route 66
204A Walsall Wood Road,
Aldridge, Midlands (W.),
WS9 8HB, UK

Tel: 0922 59398

Roy Smith Motors
116/124 Burlington Road,
New Malden, Surrey, UK

Tel: 081 949 6600

**Royal Enfield/Hitchcocks
Motorcycles**
Rosemary Cottage,
Oldwich Lane W, Chadwick
End, Solihull, Midlands
(W.), B93 OBJ, UK

Tel: 0564 783192
Fax: 0564 783313

Specialist supplier of Royal
Enfield spares, new and used.
Wide range of Amal carburettors
and spares. Mail order to
anywhere in the world. Callers
by appointment. Business hours
2pm-7pm. Weekdays only.

SP Tyres (UK) Ltd
Fort Dunlop, Erdington,
Birmingham, Midlands (W.),
B24 9QT, UK

Contact: Mr S.C.Male
Tel: 021 306 2940
Fax: 021 306 2359

The manufacturer and
distributor of Dunlop motorcycle
tyres and tubes for all
motorcycle activities including:
road, road race, moto cross, trials
and speedway.

SPC Bearings Ltd
Wood Street, Lye,
Stourbridge, Midlands (W.),
DY9 8RX, UK

Tel: 0384 896287

**Saddleworth Classic
Motorcycles**
Knarr Mill, Oldham Road,
Delph, Lancashire,
OL3 5RQ, UK

Tel: 0457 872788

Sevenhills Motorcross
Sevenhills House, Ingham,
Bury St Edmunds, Suffolk,
UK

Tel: 0284 728018

Shield Motorcycle Spares
32 Cage Hill, Swafham
Prior, Cambridgeshire,
CB5 OJS, UK

Tel: 0638 743667

AJS/Matchless second hand
spares. Tel/Fax 0638-743667.

Simpson Mecanique
2 Bis Rue de Corneilhan,
Beziers, 34500, France

**Simpson Mecanique
Ducati**
Routes de Saint Georges,
Juvignac, 34990, France

Tel: 010 33 670 30700
Fax: 010 33 670 30945

Modern Ducati dealer, has
dynamic wide-case (post '68)
single spares department with
massive stocks of racing parts
and accessories. English
catalogue available. Stocks parts
for all V-Twins too.

Slaters
Collington, nr Bromyard,
Hereford & Worcester,
NR7 4ND, UK

Tel: 0885 410295

Southern Cycles
Eastwick Road, Great
Bookham, Surrey,
KT23 4DT, UK

Tel: 0372 450050

Soviet Knight 650
Brockholme, Seaton Rd,
Hornsea, Humberside (N.),
UK

Tel: 0964 533878
Fax: 0964 534560

Spares (GB)
1 Walpole Road, London,
SW19 2BZ, UK

Tel: 081 540 7155
Fax: 081 543 5872

Ducati and Moto Guzzi spares
and accessories - Koni shock
absorbers, Ferodo brake pads.
Windscreens for all makes.
Rapid delivery specialist. Export
welcome.

Spares Unlimited
437 Hessle Road,
Hull, Humberside (N.),
HU3 4EH, UK

Tel: 0482 20385
Fax: 0482 23142

Sporting Cycles
1A Gipsy Road, West
Norwood, London,
SE27 9TD, UK

Tel: 081 670 8108

Stators
Unit 71, City Business Park,
Somerset Place, Stoke,
Plymouth, Devonshire,
PL3 4BB, UK

Tel: 0752 560906

Steve's Motorcycles
Bolney Road, Cowfold,
Horsham, Sussex (W.), UK

Tel: 040 3864 533

**Stevens & Stevens Trials
Centre**
Unit 43,Blue Chalet Ind.
Park, London Road, West
Kingsdown, Kent,
TN15 6BQ, UK

Tel: 0474 854265
Fax: 0474 854032

Trials motorcycles, spares,
accessories & clothing. Advice
on joining clubs (adults &
juniors).

Supreme Motorcycles
Hilltop, No.1 High Street,
Earl Shilton, Leicestershire,
LE9 7DP, UK

Tel: 0455 841133

T & G Motorcycles
4 The Parade, Wiltshire
Road, Thornton Heath,
Surrey, CR4 7QN, UK

Tel: 081 684 1414

T. Johnson Cables GB
5 Laburnum Grove,
Banbury, Oxfordshire,
OX16 9DP, UK

Tel: 0926 651470

TMS
92-94 Carlton Road,
Nottingham,
Nottinghamshire,
NG3 2AS, UK

Tel: 0602 503 447

TRM Racing
Terry Rudd Motorcycles,
Fen Road, Holbeach,
Spalding, Lincolnshire, UK

Tel: 0406 22430

TT Tyres
44 Jamaica Street,
Liverpool, Merseyside,
L1 0AF, UK

Tel: 051 707 1616

Taffspeed
128 Corporation Street,
Newport, Gwent, NP9
0BH, UK

Tel: 0633 840450

Taylor Racing
23-25 Station Hill,
Chippenham, Wiltshire, UK

Tel: 0249 657575/6

**Telford Motorcycle
Breakers**
49-53 High Street,
Wellington, Telford,
Shropshire, TF1 1JT, UK

Tel: 0952 248949

Telford's motorcyle breakers are
Shropshire's busiest and best
new and used motorcycle spares
specialist. Vast stocks of new and
used Japanese spares. Bikes and
spares bought for cash.

Terry Hobbs Motorcycles
2 Camden Street, Sherwell,
Plymouth, Devonshire, UK

Tel: 0752 662429

The Bonneville Shop
14 Pomeroy Close, Canley,
Coventry, Midlands (W.),
CV4 8AZ, UK

***Contact:** Humphries, Clive*
Tel: 0926 56713

**The Goodman
Engineering Co. Ltd**
Westward, Buckle St,
Honeybourne, Evesham,
Hereford & Worcester,
WR11 5QQ, UK

Tel: 0386 832090
Fax: 0386 831614

The Old Works
Dover Street, Maidstone,
Kent, UK

Tel: 0622 720430

The Tank Shop
Glenview, Glenmidge,
Auldgirth, Dumfries &
Galloway, DG2 0SW, UK

Tel: 0387 74259

Thoroughbred Stainless
Unit 9, Potts Marsh Ind.Est.,
Westham, Pevensey,
Sussex (E.), BN24 5NH, UK

Tel: 768564

Thunderbird Classics
71 Church Street, Horwich,
Bolton, Manchester, Gt.,
BL6 6AA, UK

Tel: 0204 697275

Toga
Rigby Close, Heathcote Ind.
Estate, Heathcote,
Warwickshire, CV34 6TL,
UK

Tel: 0926 497375
Fax: 0926 400807

**Tommy Robb
(Motorcycles) Ltd**
240 Manchester Road,
Warrington, Cheshire,
WA1 3BE, UK

Tel: 0925 56528
Fax: 0925 234518

Main agents for Honda,
Triumph, Kawasaki, Yamaha,
Piaggio motorcycles. We offer a
vast range of accessories:
clothing, helmets, boots, gloves,
etc.; plus advice on training,
insurance, finance etc. Our
service department is the best for
repairs and MOTs.

Tony Bairstow
5 Chelsea Manor Court,
Flood Walk, London,
SW3 5SA, UK

Tel: 071 352 3972

Harley-Davidson obsolete parts
for sale. Full stock of spares for
750cc and 1200cc from 1936 to
1964. Also available is a full
range of parts, books and
manuals. No callers please.

Triumph BSA
92-94 Carlton Road,
Nottingham,
Nottinghamshire,
NG3 2AS, UK

Tel: 0602 503447
Fax: 0602 503566

**Triumph Motorcycle
Spares**
14-18 Kew Foot Road,
Richmond, Surrey,
TW9 2SS, UK

Tel: 081 940 2045

Two Wheels
35 Cross Street,
Farnborough, Hampshire,
UK

Tel: 0252 522552

Vale - Onslow
104-116 Stratford Rd,
Sparbrook, Birmingham,
Midlands (W.), B11 1AW,
UK

Tel: 021 772 2577
Fax: 021 772 5837

Vic Nunn
113/117 Brighton Road,
Surbiton, Surrey, KT6 5NJ,
UK

Tel: 081 399 2455

Voc Spares Company
The Wharf, Burford Lane,
Lymm, Cheshire,
WA13 0SL, UK

Tel: 0925 753367

Webbs of Lincoln
117-121 Portland Street,
Lincoln, Lincolnshire,
LN5 7LG, UK

Tel: 0522 528951

West Pier Motorcycles
24/25 Brandon Terrace,
Edinburgh, Lothian,
EH3 1VZ, UK

Tel: 031 556 3286

Wilemans Motors
Siddals Road, Derby,
Derbyshire, DE1 2PZ, UK

Tel: 0332 42813
Fax: 0332 384680

Worsley Motorcycles
103a Manchester Road,
Walkden, Worsley, Man-
chester, Gt., M28 5NT, UK

Tel: 061 799 8849

Section 13

TOOLS

ARE Ltd
Woking Road, Guildford,
Surrey, GU1 1QD, UK

Tel: 0483 33163

Cetem Polishing Kits
Dept A, Freepost,
Birmingham, Midlands (W.),
M6 8BR, UK

Tel: 0268 769392

Chainmail
Orchard House,
Applelands, Wrecclesham,
Farnham, Surrey,
GU10 4TL, UK

Tel: 0252 793813

Davida (UK) Ltd
Millhouse Holt Avenue,
Moreton, Wirral,
Merseyside, L46 0SS, UK

Tel: 051 678 4656
Fax: 051 677 5398

Doug Richardson
20 South Street, South
Molton, Devonshire,
EX36 4AG, UK

Tel: 0769 574108

Dynic Sales
Bell View Cottage.
Ladbroke, Leamington Spa,
Warwickshire, CV33 0DA,
UK

Tel: 0926 814313

Eastwood Company
Unit G, Millbrook Road,
Stover Industrial Estate,
Yate, Bristol, Avon,
BS17 5PB, UK

Tel: 0454 329900
Fax: 0454 329988

Hard to find tools, paints,
chemical buffing, blasting,
spraying and equipment. Top
quality & value guaranteed.
Send for free catalogue.

Fine Thompson Ltd
Felspar Road, Amington,
Tamworth, Staffordshire,
B77 4DP, UK

**Hog & Chop
International**
48 Manchester Street,
Cleethorpes, Humberside
(S.), DN35 7QG, UK

Tel: 0472 697859

Hydrocontrol Systems
Chapel Works,
Laddingford, Maidstone,
Kent, ME18 6BT, UK

Tel: 0622 872146

**Light Soldering
Developments Ltd**
97-99 Gloucester Road,
Croydon, Surrey,
CR0 2DN, UK

Tel: 081 689 0574
Fax: 081 689 0090

Soldering irons, 12 watts to 75
watts, 6 volts to 240 volts.
Suitable for electrical, control
cable and light metal work. Mail
order list available.

M. J. Roden
Overhill Stables, Sandy
Lane, Tilford, Farnham,
Surrey, GU10 2ET, UK

Tel: 025 125 4336

Mercer Skilled Crafts Ltd
Springfield Works,
Moorside, Cleckheaton,
Yorkshire (W.), BD19 6JT,
UK

Tel: 0274 872861

Nady/Powerline Limited
Marion Place, Port Grat,
St.Sampsons, Guernsey, CI,
UK

P&D Consultants
P.O.Box 12, Oakham,
Leicestershire, LE15 8LQ,
UK

Peatol Machine Tools
19 Knightlow Road,
Harborne, Birmingham,
Midlands (W.), B17 89S, UK

Penrite Oil
Unit 1B, 31 Dollman Street,
Birmingham, Midlands (W.),
B7 4RP, UK

Renntec
69 Woolsbridge Ind. Est.,
Three Legged Cross,
Wimbourne, Dorset,
BH21 6SP, UK

Tel: 0202-826722

Rick James
1a Suggits Lane,
Cleethorpes, Humberside
(S.), DN35 7JE, UK

Tel: 0472 694468

Rotascraft Engineering
Unit 5B, Willowtree
Ind.Est, Alnwick,
Northumberland,
NE66 2PF, UK

Tel: 0665 603362

Shadowfax Engineering
Billing Station House,
Cogenhoe,
Northamptonshire,
NN7 1NQ, UK

Tel: 0604 890
Fax: 0604 890885

Specialist supplier of motorcycle
workshop equipment: range
includes hydraulic & pneumatic
motorcycle lifts, compressors,
blast cabinets, tool storage, range
of diagnostic equipment & hand
tools.

**The Electrical Parts
Company Limited**
Unit 1A, Horns Trading
Estate, Bromyard Road,
Tenbury Wells, Hereford &
Worcester, WR15 8DE,
UK

Tel: 0584 819570

**The Tuning Shop
Limited**
Dept E2, PO Box 48,
Epsom, Surrey, KT17 4BR,
UK

Tracy Tools (BB) Ltd
2 Mayors Avenue,
Dartmouth, Devonshire,
TQ6 9NC, UK

Tel: 0803 833134

Suppliers of standard and
special thread taps & dies for
restoration work. Send for free
catalogue. (Cutting tools of all
types.)

Triumph Basket Cases
2 Kemp Road, Winton,
Bournemouth, Dorset,
BH9 2PW, UK

Tel: 0202 514446

Section 14

TRAINING

Archway Project
Archway 6, Harrow Manor Way, Thamesmead, London, SE13 5RS, UK

Tel: 081 310 1730

An organisation funded on grants/sponsorship. Teaches mechanical, riding and driving skills to young people. Actively encourages participation from women and ethnic minorities.

BMF Rider Training Scheme
PO Box 2, Uckfield, Sussex (E.), TN22 3ND, UK

Tel: 0825 712896
Fax: 0825 712787

Bridge South West Motorcycle Training
Centre, Cofton Road, Exeter, Devonshire, EX2 8QW, UK

Tel: 0392 216021

Open every day for C.B.T, test preparation and advanced courses. Purpose designed centre, hire bikes, radio tuition and highly qualified instructors. Quite possibly the best!

Bristol Motorcycle Training Super Centre
Old Gloucester Road, Ham Brook, Bristol, Avon, BS16 1RS, UK

Tel: 0454 776333

C J Wilson
23-25 West Main Street, Uphall, Broxburn, Lothian, UK

Tel: 0506 856751

Cheshire Motorcycle Training
111 Middlewhich Road, Northwich, Cheshire, CW9 5BY, UK

Tel: 0606 49084
Fax: 0606-45705

Compulsory basic training, supercourses, all training to full licence. Hire bikes. 95% First time pass rate.

Donington Race School
Castle Donington, Derby, Derbyshire, DE7 2RP, UK

Tel: 0507 343 445

Mercer Skilled Crafts Ltd
Springfield Works, Moor-side, Cleckheaton, York-shire (W.), BD19 6JT, UK

Tel: 0274 872861

Shire Training Services
90 Green Leys, St Ives, Huntingdon, Cambridge-shire, PE17 4SA, UK

Tel: 0408 464689

Swarbrick Racing
The Forge, Garstang Road, Brock, Nr Preston, Lancashire, PR3 ORD, UK

Tel: 0995 40291

TT Motorcycles
Grigor Hill Ind Estate, Nairn, Highland, IV12 5HY, Scotland

Tel: 0667 53540
Fax: 0667 52837

Section 15

TUNING

Boxer-K Motorcycle Services Co
70 Hillcrest Road, Offerton. Stockport, Manchester, Gt., SK2 55E, UK

Contact: Neil Sagar
Tel: 061 483 7367

Engine tuning specialists for BMW. Bosch fuel injection systems, Jetronic, Motronic, and Bing carburettors. Latest diagnostic equipment. Spares/repairs, while you wait by appointment. Phone anytime.

Boyer Bransden Electronics Ltd
Frindsbury House, Cox Lane, Detling, Maidstone, Kent, ME14 3HE, UK

Tel: 0622 730939
Fax: 0622 730930

Contactless ignition systems with electronic advance for road and racing. Alternator regulators for 6 to 12 volt conversions and batteryless running. Plus technical help line.

Carl Rosner Ltd
Station Approach, Sanderstead Road, South Croydon, Surrey, CR2 OPL, UK

Contact: Carl Rosner
Tel: 081 657 0121
Fax: 081 651 0596

Established 25 years. Comprehensive spares stock for Commando, Triumph 650/750. Trident/R-3. Full workshop facilities, including restorations. Dealers for Hinckley Triumphs, Norton Rotary and Indian Enfield.

George Hopwood Motorcycles
112 Wren Road, Sidcup, Kent, DA14 4NF, UK

Contact: George Hopwood
Tel: 081 300 9573

Triumph specialist. Engine overhauls. Total restorations. T120 Thruxton parts - exhausts pipes, camshafts seats, fairings, etc. Full tuning services. Free advice and discounts to TOMCC members.

Motul Motor Oil (UK)
Sidings Road, Lowmoor Road Ind.Estate, Kirkby in Ashfield, Nottinghamshire, NG17 7JZ, UK

Tel: 0623 757262
Fax: 0623 757049

UK distributor of high quality road and racing synthetic lubricants and ancilliary products.

Section 16

ENDURO

Caerleon & District Motor Sports Club
14 Highfield, Goytree, Pontypool, Gwent, UK

Contact: Mr S Jones
Tel: 0873 880130

Caerphilly & DMCC
2 Ffwrwn Road, Machen, Glamorgan, Mid-, NP1 8NF, UK

Contact: Mr P Lear

Carshalton MCC
31 Preston Drive, Ewell, Epsom, Surrey, KT19 0AD, UK

Contact: Mr Martin Ireland
Tel: 081 394 1262

Castle (Colchester) MCC
3 Sproughton Court, Sproughton, Ipswich, Suffolk, IP8 3AJ, UK

Contact: Mr R Foulkes
Tel: 0473 49098

Croydon MCC
Flat 4, Nelson Court, 17 Denmark Road, Carshalton, Surrey, SM5 2JH, UK

Contact: Mr S Sharp

Derwent MCC
11 Hillside, Holloway, Matlock, Derbyshire, UK

Contact: Mrs E Mason

Diss & District MC & LCC
Five Gables, Wortham Ling, Diss, Norfolk, IF22 1ST, UK

Contact: Mrs Bavin
Tel: 0379 643960

Eboracum MC
2 Greenway, Huntington, Yorkshire (N.), YO3 9QE, UK

Contact: Mr C Cass
Tel: 0904 760384

Essex Enduro Club
37 Gazelle Drive, Canvey Island, Essex, SS8 7NB, UK

Contact: Mr N Petty
Tel: 0268 680906

Gilfach Triangle Motor Club
2 William Street, Aberbargoed, Glamorgan, Mid-, CF8 9FP, UK

Contact: Mr L Bowen

Llangollen & DMCC
2 Penllyn, Nant Parc, Johnstown, Wrexham, Clwyd, LL14 1YG, UK

Contact: Mr J Price
Tel: 0978 842142

Lowestoft Invaders MCC
58 Dell Road, Oulton Broad, Lowestoft, Suffolk, NR33 9NS, UK

Contact: Mr R Greengrass
Tel: 0502 563566

Manby Showground
Sunny Oak, Little Cawthorpe, Louth, Lincolnshire, LN11 8ND, UK

Contact: James Tointon
Tel: 0507 604375
Fax: 0507 604092

Off road practice ground with specialist tracks for quads, pilots, moto-x, enduro, trials and 4 x 4. Test sessions must be prebooked by phone Monday to Saturday.

Merthyr Motor Club
Imperial Hotel, High Street, Merthyr Tydfil, Glamorgan, Mid-, UK

Contact: Mr B Marsh

Motorcycle Seatworks
366 Woodside Road, Wyke, Bradford, Yorkshire (W.), BD12 8HT, UK

Tel: 0274 604672

Manufacturers of every type of motorcycle seat cover, specialists in moto-cross. Appointment always necessary.

Scottish Auto-Cycle Association
Block 2, Unit 6, Whiteside Ind. Estate, Bathgate, Lothian, EH48 2RX, UK

Governing body of motor cycle sport in Scotland.

Southern MCC Ltd
8 Birchleigh Close, Onchan, Isle of Man, UK

Contact: Mrs B Jones
Tel: 0624 622408

Stowmarket & DMCC
25 Quinton Road, Needham Market, Suffolk, IP6 8BP, UK

Contact: Mrs V Hearn
Tel: 0449 721042

Sudbury MCC
14 Constable Road, Sudbury, Suffolk, CO10 6UG, UK

Contact: Mr M J Edwards
Tel: 0787 77033

Thirsk MC
2 Herriott Way, Thirsk, Yorkshire (N.), UK

Contact: Mr A Kendal
Tel: 0845 725937

Woodbridge & DMCC
8 Aldeburgh Road, Leiston, Suffolk, IP16 4ED, UK

Contact: Mrs Doreen Burrell
Tel: 0728 830371

Section 17

GRASS TRACK

500cc Sidecar Association
64 Sycamore Road, Stowupland, Suffolk, IP14 4DR, UK

Contact: Dave Brown
Tel: 0449-675184

National sidecar association for 500cc sidecar grasstracks racing (the International Formula). Promotion and information provided for the class and the association members.

Antelope MCC (Coventry) Ltd
29 Cherry Lane, Hampton Magna, Warwick, Warwickshire, CV35 8SL, UK

Contact: Mr A H Davies

Ashford MCC
207 Canterbury Road, Kennington, Ashford, Kent, TN24 9QH, UK

Contact: Roger Rigg
Tel: 0233 634271

Astra MCC
71 West Street, Sittingbourne, Kent, ME10 1AN, UK

Contact: Mr W Chesson
Tel: 0795 472926

Bantam Grasstrack Association
10 Duchess Drive, Bridgnorth, Shropshire, UK

Contact: Mr P Bearman
Tel: 07462 4587

Bewdley MCC
12 Lenchville, Broadwaters, Kidderminster, Hereford & Worcester, DY10 2YU, UK

Contact: Mrs H Taylor

Braintree & DMCC
34 Walter Way, Silver End, Witham, Essex, CM8 3RJ, UK

Contact: Mrs S Young
Tel: 0376 83045

Brighton & District MCC
73 Eastbrook Road, Portslade, Sussex (E.), BN41 1P3, UK

Contact: Ian Swyer

Brighton & District MCC is the longest established club on the South coast. Meets at the Southern Cross club, Victoria Road, Portslade, East Sussex. All social and sports activities catered for. BMF and ACU affiliated. Contact Ian Swyer on 0273 430458.

British Sporting Sidecar Association
The Mill, Bodilly, Helston, Cornwall, UK

Contact: Mr E J Seymour
Tel: 0326 574925

Chelmsford & DAC
3 Eaton Way, Great Totham, Malden, Essex, CM9 8EE, UK

Contact: Mrs H Gulliver
Tel: 0621 892606

Cornwall Solo Grass Track Club
8 Creeping Lane, The Lidden, Penzance, Cornwall, UK

Contact: Mrs C Jarvis
Tel: 0736 63616

Crawley & DMCC
Rowan, Bonehurst Road, Horley, Surrey, RH6 8QG, UK

Contact: Mrs S Slight
Tel: 0293 775798

Dunmow & DMCC Ltd
6 Beaumont Hill, Great Dunmow, Essex, CM6 2AP, UK

Contact: Mr Julian Sayer
Tel: 0371 874756

Affiliated to Eastern centre of The Auto-Cycle Union. Club caters for adult off-road events, particularly grass track, moto-cross and trials. Events organised for 1994 include: Scramble - Stebbing, Essex 01/05/94. Grass track - Ugley, Stansted, Essex 19/06/94 & 18/09/94. Membership enquiries to Dean Lambert, Tel:0245-441072.

East Coast Off Road Club
Morven, Westbourne Road, Hornsea, Humberside (N.), HU18 1PQ, UK

Contact: Mrs P Wood
Tel: 0964 533321

Eastern Sporting Sidecar Association
Cregny-Baa, 88 Glebe Road, Kelvedon, Essex, CO5 9JS, UK

Contact: Mrs Y Smith
Tel: 0376 70970

Evesham MCC
Honey Dew, Plough Road, Tibberton, Hereford & Worcester, WR9 7NQ, UK

Contact: Mr D Kimberley
Tel: 090 56395

Leamington Victory MC & LCC
12 Dadglow Road, Bishops Itchingham, Warwickshire, CV33 OTG, UK

Contact: Mr A Halford
Tel: 0926 613202

Midland Grass Track Club
7 Roseland Road, Kenilworth, Warwickshire, CV8 1GA, UK

Contact: Janet Jones
Tel: 0926 57025

Norfolk & Suffolk Jnr MCC
Mill House, Stone Street, Crowfield, Ipswich, Suffolk, IP6 9SZ, UK

Contact: Mrs J Gibbons
Tel: 044 979 397

Northallerton & DMC
6 Wensley Road, Romanby, Northallerton, Yorkshire (N.), UK

Contact: Mrs Y Kirk
Tel: 0609 780105

Northwich & Knutsford MCC
65 Green Park, Weaverham, Northwich, Cheshire, CW8 3EH, UK

Contact: Mrs L Horrigan
Tel: 0606 853093

Pendennis MC & LCC
38 Lanner Hill, Lanner, Redruth, Cornwall, UK

Contact: Mrs S Pooley
Tel: 0209 215543

Pickering & District Motor Club
4 Brockfield Road, Huntingdon Road, York, Yorkshire (N.), YO3 9DZ, UK

Contact: Mr D A Brown
Tel: 0904 622274

Pickering and District Motor Club promotes motorcycle grass track, motocross and trials. The club has been in existence since 1950.

Point of Ayr GTC
78 Twinnies Road,
Wilmslow, Cheshire, SK9
4BP, UK

Contact: Mr R Carter
Tel: 0625 527288

**Scottish Auto-Cycle
Association**
Block 2, Unit 6, Whiteside
Ind. Estate, Bathgate,
Lothian, EH48 2RX, UK

Governing body of motor cycle
sport in Scotland.

**Severn Valley Grass
Track Club**
14 St Davids Close, Lickhill
Road North,
Stourport-on-Severn,
Hereford & Worcester,
DY13 8RZ, UK

Contact: Mr J Doughty
Tel: 02993 4922

**Shropshire Grass Track
Club**
38 Monkmoor Avenue,
Monkmoor, Shrewsbury,
Shropshire, SY2 5EB, UK

Contact: Mr B Curran
Tel: 0743 58226

Skegness & DMCC
Tel e Not, Eudykes,
Friskney, Boston,
Lincolnshire, PE22 8RU, UK

Contact: Mrs P Ingham
Tel: 0754 84 350

**Spalding & Tongue End
AC**
45 Tanglewood,
Werrington, Peterborough,
Cambridgeshire, PE4 5DH,
UK

Contact: Mrs S Cotton
Tel: 0733 74249

**Stourbourne (Haverhill)
MCC**
79 Nothfield Park, Soham,
Cambridgeshire, UK

Contact: Mrs K Edson
Tel: 0353 722681

Venhill Engineering Ltd
21 Ranmore Road,
Dorking, Surrey, RH4 1HE,
UK

Tel: 0306 885111
Fax: 0306-740535

Manufacturers of Nylocable and
Featherlite control cables,
Powerhose high performance
stainless braided brake hoses.
Distributors of Magura controls,
Ariete and Buzzetti accessories
and titanium and aluminium
fasteners.

**Wainfleet & District
Sporting MCC**
New Farm, Wainfleet Bank,
Wainfleet St Mary,
Skegness, Lincolnshire,
PE24 4JU, UK

Contact: Mrs S Blackbourn

Whitchurch MCC
6 Chapel Crescent,
Darliston, Whitchurch,
Shropshire, UK

Contact: Mr C G Mellor
Tel: 0948 840143

Woodbridge & DMCC
8 Aldeburgh Road, Leiston,
Suffolk, IP16 4ED, UK

Contact: Mrs Doreen Burrell
Tel: 0728 830371

Wymondham & DMC
10 Malton Court,
Hethersett, Norwich,
Norfolk, NR9 3HE, UK

Tel: 0603 810934

HILL CLIMB

**Andreas Racing
Association**
51a Victoria Street,
Douglas, Isle of Man, UK

Contact: Mr E N Bowers
Tel: 0642 623731

**British Motor Cycle
Owners Club**
Tighsona, Rosevallon Lane,
Townend, Bodmin,
Cornwall, UK

Contact: Mrs S Stephens
Tel: 0208 77940

Pendennis MC & LCC
38 Lanner Hill, Lanner,
Redruth, Cornwall, UK

Contact: Mrs S Pooley
Tel: 0209 215543

Welsh Sprint Society
22 Cedar Road, Trinant,
Crumlin, Gwent, UK

Contact: Mrs S Pope
Tel: 0495 214895

West Cornwall MC
Oakland Cottage, Paul,
Penzance, Cornwall, UK

Contact: Mr B Ellis
Tel: 073 673529

MOTO CROSS

Astra MCC
71 West Street,
Sittingbourne, Kent,
ME10 1AN, UK

Contact: Mr W Chesson
Tel: 0795 472926

Barham DMC & LCC
16 Hog Grenn, Elham,
Canterbury, Kent,
CT4 6TU, UK

Contact: Mr E G Mears
Tel: 0303 840435

Bewdley MCC
12 Lenchville, Broadwaters,
Kidderminster, Hereford &
Worcester, DY10 2YU, UK

Contact: Mrs H Taylor

Bognor Regis & DMCC
"Wayside", North
Mundham Farm,
Chichester, Sussex (W.),
PO20 6JY, UK

Contact: Mr A Martin
Tel: 0243 785697

Boltby Moto Cross Club
26 Kings Meadow,
Sowerby, Thirsk, Yorkshire
(N.), YO7 1PA, UK

Contact: Mr B Moody
Tel: 0845 523463

Braintree & DMCC
34 Walter Way, Silver End,
Witham, Essex, CM8 3RJ,
UK

Contact: Mrs S Young
Tel: 0376 83045

Brighton & District MCC
73 Eastbrook Road,
Portslade, Sussex (E.),
BN41 1P3, UK

Contact: Ian Swyer

Brighton & District MCC is the
longest established club on the
South coast. Meets at the
Southern Cross club, Victoria
Road, Portslade, East Sussex. All
social and sports activities
catered for. BMF and ACU
affiliated. Contact Ian Swyer on
0273 430458.

**Bury St Edmunds &
DMCC**
Glastonbury House, 6
Westminster Drive, Bury St
Edmunds, Suffolk,
IP33 2EZ, UK

Contact: Mrs R J Bowers
Tel: 0284 761828

**Cambourne-Redruth &
DMC & LCC**
7 Moresk Close, Truro,
Cornwall, UK

Contact: Mr B Jennings
Tel: 0872 78317

Castle (Colchester) MCC
3 Sproughton Court,
Sproughton, Ipswich,
Suffolk, IP8 3AJ, UK

Contact: Mr R Foulkes
Tel: 0473 49098

Chelmsford & DAC
3 Eaton Way, Great
Totham, Malden, Essex,
CM9 8EE, UK

Contact: Mrs H Gulliver
Tel: 0621 892606

**Cheshire North West
Schoolboys SC**
Stud Cottage, Farm Road,
Oakmere, Northwich,
Cheshire, CW8 2HD, UK

Contact: Mrs S Alexander
Tel: 0606 883613

**Cornwall Moto Cross
Club**
1 Praze Meadow, Penryn,
Cornwall, UK

Contact: Mrs B Warmington
Tel: 0326 74723

Coventry Junior MCC
Woodside Farm, Hardbury
Lane, Bishops Tachbrook,
Leamington Spa,
Warwickshire, UK

Contact: Mr J Tomson

Crawley & DMCC
Rowan, Bonehurst Road,
Horley, Surrey, RH6 8QG,
UK

Contact: Mrs S Slight
Tel: 0293 775798

Derby Phoenix MCC
1 The Hill, Swarkstone
Road, Chellaston,
Derbyshire, DE73 1UD, UK

Contact: Mr D Smith
Tel: 0332 705885

Derwent MCC
11 Hillside, Holloway,
Matlock, Derbyshire, UK

Contact: Mrs E Mason

**Diss & District MC &
LCC**
Five Gables, Wortham Ling,
Diss, Norfolk, IP22 1ST,
UK

Contact: Mrs Bavin
Tel: 0379 643960

Dragon Schoolboys SC
70 Nantwich Road,
Middlewich, Cheshire,
CW10 9HG, UK

Contact: Mr A Hughes
Tel: 0606 843266

Dunmow & DMCC Ltd
6 Beaumont Hill, Great
Dunmow, Essex,
CM6 2AP, UK

Contact: Mr Julian Sayer
Tel: 0371 874756

Affiliated to Eastern centre of
The Auto-Cycle Union. Club
caters for adult off-road events,
particularly grass track,
moto-cross and trials. Events
organised for 1994 include:
Scramble - Stebbing, Essex
01/05/94. Grass track - Ugley,
Stansted, Essex 19/06/94 &
18/09/94. Membership enquiries
to Dean Lambert,
Tel:0245-441072.

**East Anglian Schoolboy
Scrambling Club**
142 Blackshots Lane, Grays,
Essex, RM16 2LH, UK

Contact: Mrs C Hornsell
Tel: 0375 378234

**East Wales Moto Cross
Club**
4 The Bungalows,
Heol-Fawr, Nelson,
Glamorgan, Mid-,
CF46 6NR, UK

Contact: Mr R Mantel
Tel: 0443 450401

**Essex & Suffolk Border
MCC**
9 The Crescent, Barham,
Ipswich, Suffolk, IP6 0PE,
UK

Contact: Mr P Fenn
Tel: 0473 830666

Harwich MCC
28 Lodge Road, Little
Oakley, Harwich, Essex,
CO12 5ED, UK

Contact: Mr P Goodwin
Tel: 0255 886004

Hull & DMC
Central Farm, Skirpenbeck,
York, Yorkshire (N.),
YO4 1HF, UK

Contact: Mrs P Stannard
Tel: 0759 71984

Ipswich MC & LCC
9 St Mary's Close,
Bramford, Ipswich, Suffolk,
IP8 4DL, UK

Contact: Mrs J Sago
Tel: 0473 742468

**Isle Of Man Schoolboy
Motor Cycling**
Ltd, Min-y-don,
Ballaquayne Park, Isle of
Man, UK

Contact: Mr Kneen
Tel: 0624 842582

Kings Lynn & District MX Club
13 Blenheim Road, Kings Lynn, Norfolk, PE30 3HE, UK

Contact: Mr R Reeve
Tel: 0553 675406

Langbaurgh DMXC
West Banks Farm, Glaisdale, Whitby Yorkshire (N.), YO21 2QP, UK

Contact: Mrs C Webster
Tel: 0947 87359

Littleport & DMC & LCC
14 North Street, Freckenham, Bury St Edmunds, Suffolk, IP28 8HY, UK

Contact: Mr P Gammon
Tel: 0638 721046

Louth & DMCC
25 Tudor Drive, Louth, Lincolnshire, LN11 9EE, UK

Contact: Mr Dunham
Tel: 0507 607518

Merthyr Motor Club
Imperial Hotel, High Street, Merthyr Tydfil, Glamorgan, Mid-, UK

Contact: Mr B Marsh

Motorcycle Seatworks
366 Woodside Road, Wyke, Bradford, Yorkshire (W.), BD12 8HT, UK

Tel: 0274 604672

Manufacturers of every type of motorcycle seat cover, specialists in moto-cross. Appointment always necessary.

Nantwich & DMC
20 Cawfield Avenue, Widnes, Cheshire, UK

Contact: Mrs M McMinn
Tel: 051 423 1098

National Motorcycle Centre Trials Club
Park Promotions Ltd, The Post House, Ham Hill, Powick, Hereford & Worcester, WR2 4RD, UK

Contact: Mr T Matthews
Tel: 0905 830459

Newport & Gwent Motor Club
7 Linden Grove, Rumney, Cardiff, UK

Contact: Mr W Aston

Norfolk & Suffolk Jnr MCC
Mill House, Stone Street, Crowfield, Ipswich, Suffolk, IP6 9SZ, UK

Contact: Mrs J Gibbons
Tel: 044 979 397

North Weald MCC
Willow Cottage, 10 St Marys Avenue, Wanstead, London, E11 2NP, UK

Contact: Mr K Wilson
Tel: 081 989 5046

Northallerton & DMC
6 Wensley Road, Romanby, Northallerton, Yorkshire (N.), UK

Contact: Mrs Y Kirk
Tel: 0609 780105

Norwich & Dist SB/AD MXC
90 Melton Road, Wymondham, Norfolk, NR18 0DE, UK

Contact: Mrs E Clark
Tel: 0953 603451

Norwich Viking MCC
10 Willow Close, Lingwood, Norwich, Norfolk, NR13 4BT, UK

Contact: Mr A Hay
Tel: 0603 713017

Phil Ayliff Products Ltd
25 Alliance Close, Attleborough Fields, Nuneaton, Warwickshire, CV11 6SD, UK

Contact: Trevor Ayliff
Tel: 0203 343741
Fax: 0203 641247

Manufacturer of Dunlopad sintered metal brake pads.

Pickering & District Motor Club
4 Brockfield Road, Huntingdon Road, York, Yorkshire (N.), YO3 9DZ, UK

Contact: Mr D A Brown
Tel: 0904 622274

Pickering and District Motor Club promotes motorcycle grass track, motocross and trials. The club has been in existence since 1950.

Salop MC
8 Oakwood Drive, Heath Farm, Shrewsbury, Shropshire, SY1 3EA, UK

Contact: Mr A Johnson

Sandwell Heathens
49 Valley Road, Lye, Stourbridge, Midlands (W.), DY9 8JG, UK

Contact: Mr N Skidmore

Scarborough MC
20 Broadlands Drive, East Ayton, Scarborough, Yorkshire (N.), YO13 9ET, UK

Contact: Mrs E Race
Tel: 0723 863987

Scottish Auto-Cycle Association
Block 2, Unit 6, Whiteside Ind. Estate, Bathgate, Lothian, EH48 2RX, UK

Governing body of motor cycle sport in Scotland.

South Humberside Youth MCC
15 Forest Way, Humberston, Grimsby, Humberside (S.), DN36 4HQ, UK

Contact: Mrs C Dolby
Tel: 0724 815806

Southend & DMCC
Rosebank, Softwater Lane, Hadleigh, Benfleet, Essex, SS7 2NE, UK

Contact: Mrs M Spurgeon
Tel: 0702 554830

Southern MCC Ltd
8 Birchleigh Close, Onchan, Isle of Man, UK

Contact: Mrs B Jones
Tel: 0624 622408

Stourbourne (Haverhill) MCC
79 Nothfield Park, Soham, Cambridgeshire, UK

Contact: Mrs K Edson
Tel: 0353 722681

Thirsk MC
2 Herriott Way, Thirsk, Yorkshire (N.), UK

Contact: Mr A Kendal
Tel: 0845 725937

Triangle (Ipswich MCC)
18 Monton Rise, Ipswich, Suffolk, IP2 9QQ, UK

Contact: Mr R Sayer
Tel: 0473 687698

Venhill Engineering Ltd
21 Ranmore Road, Dorking, Surrey, RH4 1HE, UK

Tel: 0306 885111
Fax: 0306-740535

Manufacturers of Nylocable and Featherlite control cables, Powerhose high performance stainless braided brake hoses. Distributors of Magura controls, Ariete and Buzzetti accessories and titanium and aluminium fasteners.

West Cornwall MC
Oakland Cottage, Paul, Penzance, Cornwall, UK

Contact: Mr B Ellis
Tel: 073 673529

Woodbridge & DMCC
8 Aldeburgh Road, Leiston, Suffolk, IP16 4ED, UK

Contact: Mrs Doreen Burrell
Tel: 0728 830371

QUADS

Aberaman & DMCC
28 Dare Road, Cwmdare, Aberdare, Powys, CF44 8UB, UK

Contact: Mr G Butt
Tel: 0685 881907

Andreas Racing Association
51a Victoria Street, Douglas, Isle of Man, UK

Contact: Mr E N Bowers

Tel: 0642 623731

Manby Showground
Sunny Oak, Little Cawthorpe, Louth, Lincolnshire, LN11 8ND, UK

Contact: James Tointon
Tel: 0507 604375
Fax: 0507 604092

Off road practice ground with specialist tracks for quads, pilots, moto-x, enduro, trials and 4 x 4. Test sessions must be prebooked by phone Monday to Saturday.

Motorcycle Seatworks
366 Woodside Road, Wyke, Bradford, Yorkshire (W.), BD12 8HT, UK

Tel: 0274 604672

Manufacturers of every type of motorcycle seat cover, specialists in moto-cross. Appointment always necessary.

Phil Ayliff Products Ltd
25 Alliance Close, Attleborough Fields, Nuneaton, Warwickshire, CV11 6SD, UK

Contact: Trevor Ayliff
Tel: 0203 343741
Fax: 0203 641247

Manufacturer of Dunlopad sintered metal brake pads.

Scottish Auto-Cycle Association
Block 2, Unit 6, Whiteside Ind. Estate, Bathgate, Lothian, EH48 2RX, UK

Governing body of motor cycle sport in Scotland.

Weardale Off Road Centre
Coves House Farm, Wolsingham, Co.Durham, DL13 3BG, UK

Contact: Anthony Todd
Tel: 0388 527375
Fax: 0388 526157

Bring your own machine or hire our quads or Pilots. Four excellent tracks to choose from. Ample parking. Toilets. Refreshments. Pressure washer.

ROAD RACING

Association of Motor Racing Circuit
Owners Brands Hatch Leisure plc, Fawkham Longfield, Kent, DA3 8NG, UK

Contact: Elizabeth John
Tel: 0474 872331
Fax: 0474 874766

The Association represents the interests of the UK road race circuit owners with all interested parties and works to ensure the sport's future success.

Astra MCC
71 West Street, Sittingbourne, Kent, ME10 1AN, UK

Contact: Mr W Chesson
Tel: 0795 472926

Auto 66 Club
3 New Road, Nafferton, Driffield, Humberside, YO25 0JP, UK

Contact: Mr P Hillaby
Tel: 0377 44727

Avon Park International Racing Assoc.
27 Hillfields, Foxton, Cambridgeshire, CB2 6RP, UK

Contact: Mr Talbot

Clubmans Racing Club (EM)
8 Beardsley Road, Lakeside, Quorn, Leicestershire, UK

Contact: Mr P King
Tel: 0509 416369

Darley Moor MRRC Ltd
Anvil House, Derby Road, Old Tupton, Chesterfield, Derbyshire, S42 6LA, UK

Contact: Mr E Nelson

Derby Phoenix MCC
1 The Hill, Swarkstone Road, Chellaston, Derbyshire, DE73 1UD, UK

Contact: Mr D Smith
Tel: 0332 705885

Donnington Supporters Club
80 Baldocks Lane, Melton Mowbray, Leicestershire, LE13 1EW, UK

Contact: Tony Gamble
Tel: 0664 64042

Club benefits include gate concessions, free paddock transfer, private enclosure with covered grandstand and portakabin, newsletters, circuit rides. Membership costs just £11 single, £17 joint.

Early Stocks RC
89 Mercer Avenue, Waterorton, Birmingham, Midlands (W.), B46 1NG, UK

Contact: Mr R Hooper
Tel: 021 747 7255

Early Stocks race club provides a cheap way of racing, for twin shock, air cooled machines from 1976 to present. Phone or write for more details.

East Midlands Racing Association
54 Monks Road, Binley Woods, Coventry, Midlands (W.), CV3 2BS, UK

Contact: Mr M.Jessup
Tel: 0203 544463

Forgotten Racing Club
36 Farndon Road, Sutton in Ashfield, Nottinghamshire, NE17 5HT, UK

Contact: Mr J Miller
Tel: 0623 514253

MW Leathers
18 Barking Road, Imperial Mews, London, E6 3BP, UK

Contact: *Mike Wills*
Tel: 081 471 3933
Fax: 081 471 3933

Motorcycle leather clothing for sport and road. Standard or made to measure mail order available. Visa or Access accepted. Ring for appointment.

MMCRC
Coton Post Office, Newport Road, Coton, Staffordshire, UK

Contact: *Mr A Innamorati*

Manx MCC Ltd
The Grandstand, Douglas, Isle of Man, UK

Contact: *Caroline Etherington*
Tel: 0624 627979
Fax: 0624 661923

Fortnight of road racing held on famous TT mountain course, including newcomers and classic races. Many good vantage points and grandstand seating for nominal charge.

Nantwich & DMC
20 Cawfield Avenue, Widnes, Cheshire, UK

Contact: *Mrs M McMinn*
Tel: 051 423 1098

National Scooter Sport Association
PO Box 32, Mansfield, Nottinghamshire, NG19 OAZ, UK

Contact: *J.Smith*
Tel: 0623 651658

The N.S.S.A is the sporting subsidiary to the N.S.R.A (listed under clubs/societies) and is ACU affiliated, as a non-territorial club. The N.S.S.A. is involved in all aspects of scooter sport, i.e. road racing, sprint, grass tracking.

New Era MCC
107 Mill Studio Business, Centre, Crane Mead, Ware, Hertfordshire, SE12 9PY, UK

Contact: *Jean Maslin*
Tel: 0920 444205
Fax: 0920 468686

The largest and most popular motor cycle road racing club in Britain. Organising some eighty meetings each year on all the major circuits.

Newmarket MCC
176 Studlands Park, Newmarket, Suffolk, CB8 7AR, UK

Contact: *Mr D Buckley*
Tel: 0638 660361

North East Motorcycle Racing Club
4 Hayton Close, Eastfield Glade, Cramlington, Northumberland, NE23 9FN, UK

Contact: *Ken Murray*
Tel: 0670 735235

Organisers of motorcycle road racing events mainly in the North East of England. Meet every Wednesday evening at Barley Mow, P.H.Gateshead. New members welcome.

Pegasus (Grantham) MC & CC Ltd
72 Barrowby Road, Grantham, Lincon, NG31 8AF, UK

Contact: *Mr T Harris*
Tel: 0476 65311

Phil Ayliff Products Ltd
25 Alliance Close, Attleborough Fields, Nuneaton, Warwickshire, CV11 6SD, UK

Contact: *Trevor Ayliff*
Tel: 0203 343741
Fax: 0203 641247

Manufacturer of Dunlopad sintered metal brake pads.

Replica Fairings
Unit 3E Aston Ind.Estate, Bulkington Road, Bedworth, Warwickshire, CV12 9DN, UK

Tel: 0203 311888
Fax: 0203 310262

The leading UK manufacturer and supplier of motorcycle bodywork. Massive range of road, touring, sports, half, race and twin headlamp fairings, screens, mudguards, seat panels and convertors, 1964-1994.

Retford & DMCC Ltd
4 West Carr Road, Retford, Nottinghamshire, DN22 7NN, UK

Contact: *Mr A Lane*
Tel: 0777 701103

Scottish Auto-Cycle Association
Block 2, Unit 6, Whiteside Ind. Estate, Bathgate, Lothian, EH48 2RX, UK

Governing body of motor cycle sport in Scotland.

Single Minded Ltd
Smithy Cottage, Liverpool Road, Bickerstaffe, Lancashire, L39 0EF, UK

Tel: 0695 42362
Fax: 0695 424096

We have the parts for all single cylinder/supermono riders.

Southern 100 MCR Ltd
Ellerslie Nurseries, Castletown, Isle Of Man, UK

Contact: *Mr G Peach*
Tel: 0624 822546

TGA Racing Services
Smithy Cottage, Liverpool Road, Bickerstaffe, Lancashire, L39 0EF, UK

Tel: 0695 423621
Fax: 0695 421490

We stock everything needed by the serious classic racer. We have a vehicle at race meetings and also operate by mail order worldwide.

Trident Engineering
343A Rayners Lane, Pinner, Middlesex, HA5 5EN, UK

Tel: 081 868 1476

Repairs, parts, tuning and race preparation. Full workshop facilities including milling, turning, cylinder head and brake disc skimming, helicoils etc. Backed by T.T. winning experience.

Triumph Owners MCC (Epping Forest Branch)
11 Tudor Villa, Burton Lane, Goffs Oak, Hertfordshire, EN7 6SF, UK

Contact: *Mr E Mills*
Tel: 0707 875372

Velocette Owners
24 Hallfields, Edwalton, Nottinghamshire, NG12 4AA, UK

Contact: *Mr G Bloor*

Venhill Engineering Ltd
21 Ranmore Road, Dorking, Surrey, RH4 1HE, UK

Tel: 0306 885111
Fax: 0306-740535

Manufacturers of Nylocable and Featherlite control cables, Powerhose high performance stainless braided brake hoses. Distributors of Magura controls, Ariete and Buzzetti accessories and titanium and aluminium fasteners.

TRIALS

Aberaman & DMCC
28 Dare Road, Cwmdare,
Aberdare, Powys,
CF44 8UB, UK

Contact: Mr G Butt
Tel: 0685 881907

Abergavenny Auto Club
14 Avenue Crescent,
Abergavenny, Gwent, UK

Contact: Mr D Roberts

**Antelope MCC
(Coventry) Ltd**
29 Cherry Lane, Hampton
Magna, Warwick, War-
wickshire, CV35 8SL, UK

Contact: Mr A H Davies

Ashford MCC
207 Canterbury Road,
Kennington, Ashford, Kent,
TN24 9QH, UK

Contact: Roger Rigg
Tel: 0233 634271

BSSA (East Midland)
3 Paignton Close, Aspley,
Nottingham,
Nottinghamshire,
NG8 5NX, UK

Contact: Mr A Sinclair
Tel: 0602 702902

BVM Trialsport
16 Slad Road, Stroud,
Gloucestershire,
G25 1QW, UK

Tel: 0453 762743
Fax: 0453 753972

Motorcycle trials specialists,
parts, sales, service, mail order,
access, visa. Simply the best.

Braintree & DMCC
34 Walter Way, Silver End,
Witham, Essex, CM8 3RJ,
UK

Contact: Mrs S Young
Tel: 0376 83045

Barham DMC & LCC
16 Hog Grenn, Elham,
Canterbury, Kent,
CT4 6TU, UK

Contact: Mr E G Mears
Tel: 0303 840435

Bewdley MCC
12 Lenchville, Broadwaters,
Kidderminster, Hereford &
Worcester, DY10 2YU, UK

Contact: Mrs H Taylor

Bexleyheath DMC
10 Penshurst Road,
Bexleyheath, Kent,
DA7 5ES, UK

Contact: Mr N Fleet
Tel: 0322 446556

Bognor Regis & DMCC
"Wayside", North
Mundham Farm,
Chichester, Sussex (W.),
PO20 6JY, UK

Contact: Mr A Martin
Tel: 0243 785697

Border MCC
3 Village Court, Laleham
Road, Shepperton,
Middlesex, TW17 8EQ, UK

Contact: Mr A S Avis
Tel: 0932 229990

Brighton & District MCC
73 Eastbrook Road,
Portslade, Sussex (E.),
BN41 1P3, UK

Contact: Ian Swyer

Brighton & District MCC is the
longest established club on the
South coast. Meets at the
Southern Cross club, Victoria
Road, Portslade, East Sussex. All
social and sports activities
catered for. BMF and ACU
affiliated. Contact Ian Swyer on
0273 430458.

**British Sporting Sidecar
Association**
2 New Street, Petworth,
Sussex (W.), GU28 0AS,
UK

Contact: Mr P Pesterfield
Tel: 0798 42372

**British Sporting Sidecar
Association**
7 Hall Street, Blackwood,
Gwent, UK

Contact: Mr H Lewis
Tel: 0495 222798

CSMA (South Eastern)
41 Coleridge Way,
Orpington, Kent,
BR6 0UQ, UK

Contact: Mr P Grohmann
Tel: 0689 874409

Caerphilly & DMCC
2 Ffwrwn Road, Machen,
Glamorgan, Mid-, NP1 8NF,
UK

Contact: Mr P Lear

**Cambourne-Redruth &
DMC & LCC**
7 Moresk Close, Truro,
Cornwall, UK

Contact: Mr B Jennings
Tel: 0872 78317

Cardiff Motor Club
50 Wentlog Close,
Rumney, Cardiff,
Glamorgan (S.), CF3 8HB,
UK

Contact: Mr W Cope
Tel: 0222 777279

Carshalton MCC
31 Preston Drive, Ewell,
Epsom, Surrey,
KT19 0AD, UK

Contact: Mr Martin Ireland
Tel: 081 394 1262

Castle (Colchester) MCC
3 Sproughton Court,
Sproughton, Ipswich,
Suffolk, IP8 3AJ, UK

Contact: Mr R Foulkes
Tel: 0473 49098

Cheadle (Staffs) AC
39 Parklands Road, Upper
Tean, Staffordshire,
ST10 4DT, UK

Contact: Mr P Beardmore
Tel: 0538 723232

Chelmsford & DAC
3 Eaton Way, Great
Totham, Malden, Essex,
CM9 8EE, UK

Contact: Mrs H Gulliver
Tel: 0621 892606

**Cheshire Youth Trials
Club**
4 Berwyn Road, Liscard,
Merseyside, L44 1BH, UK

Contact: Mrs S Shacklady
Tel: 051 6391904

Colwyn MCC
21 Bryn Celyn, Colwyn
Heights, Colwyn Bay,
Clwyd, UK

Contact: Mr G Taylor
Tel: 0492 533729

Conway & DMCC
Bron-Y-Gaer, Sychnant Pass
Road, Conwy, Gwynedd,
LL32 8NS, UK

Contact: Mr Arfon Jones

County Border MCC
5 Grosvenor Way, Quarry
Bank, Brierley Hill,
Midlands (W.), UK

Contact: Mr P Hipkin

Crawley & DMCC
Rowan, Bonehurst Road,
Horley, Surrey, RH6 8QG,
UK

Contact: Mrs S Slight
Tel: 0293 775798

**Crewe & South Cheshire
MC**
23 Heathfield Square,
Knutsford, Cheshire,
WA16 0AD, UK

Contact: Mrs J Clark
Tel: 0565 53677

Croydon MCC
Flat 4, Nelson Court, 17
Denmark Road, Carshalton,
Surrey, SM5 2JH, UK

Contact: Mr S Sharp

Denbigh & Mold MCC
47 Erw Goch, Ruthin,
Clwyd, LL15 1RS, UK

Contact: Mr E Clwyd Roberts
Tel: 0824 703350

Derby Phoenix MCC
1 The Hill, Swarkstone
Road, Chellaston,
Derbyshire, DE73 UD, UK

Contact: Mr D Smith
Tel: 0332 705885

Diss & District MC & LCC
Five Gables, Wortham Ling,
Diss, Norfolk, IP22 1ST, UK

Contact: Mrs Bavin
Tel: 0379 643960

Dudley & District MCC
10A Brook Road,
Oldwinsford, Stourbridge,
Midlands (W.), DY8 1NN,
UK

Contact: Mr M Atkinson
Tel: 0384 376400

Dunmow & DMCC Ltd
6 Beaumont Hill, Great
Dunmow, Essex,
CM6 2AP, UK

Contact: Mr Julian Sayer
Tel: 0371 874756

Affiliated to Eastern centre of
The Auto-Cycle Union. Club
caters for adult off-road events,
particularly grass track,
moto-cross and trials Events
organised for 1994 include:
Scramble - Stebbing, Essex
01/05/94. Grass track - Ugley,
Stansted, Essex 19/06/94 &
18/09/94. Membership enquiries
to Dean Lambert,
Tel:0245-441072.

Earl Shilton Trials Club
4b Deveron Way, Hinckley,
Leicester, Leicestershire,
LE10 0XD, UK

Contact: Mr Burdett
Tel: 0455 637304

Eastern Four Stroke Association
96 Wickham Road,
Colchester, Essex,
CO3 3EE, UK

Contact: Mr A Appleton
Tel: 0206 563579

Ebbw Vale MCC
40 Brynheulog Street, Ebbw
Vale, Gwent, UK

Contact: Mr C Barnard
Tel: 0495 304179

Eboracum MC
2 Greenway, Huntington,
Yorkshire (N.), YO3 9QE,
UK

Contact: Mr C Cass
Tel: 0904 760384

Essex & Suffolk Border MCC
9 The Crescent, Barham,
Ipswich, Suffolk, IP6 0PE,
UK

Contact: Mr P Fenn
Tel: 0473 830666

Essex Schoolboy/Adult Trials Club
7 Dryden Close, Hainault,
Ilford, Essex, IG6 3DZ, UK

Contact: Mrs R E Rouse
Tel: 081 500 3346

Evesham MCC
Honey Dew, Plough Road,
Tibberton, Hereford &
Worcester, WR9 7NQ, UK

Contact: Mr D Kimberley
Tel: 090 56395

Gilfach Triangle Motor Club
2 William Street,
Aberbargoed, Glamorgan,
Mid-, CF8 9FP, UK

Contact: Mr L Bowen

Guisborough & DMC Ltd
10 Whitby Avenue,
Guisborough, Cleveland,
TS14 7AP, UK

Contact: Mrs B Atkinson
Tel: 0287 635107

Gwent Schoolboy Trials Club
Colbost, Newport Road,
Magor, Gwent, NP6 3BZ,
UK

Contact: Mr S Payne
Tel: 0633 880442

Harwich MCC
28 Lodge Road, Little
Oakley, Harwich, Essex,
CO12 5ED, UK

Contact: Mr P Goodwin
Tel: 0255 886004

Hull AC
5 Fairholme Lane, Wawne,
Hull, Humberside (N.),
HU7 2XB, UK

Contact: Mr C Watson
Tel: 0482 822267

Humberside Hawks MCC
61 High Street,
Messingham, Scunthorpe,
Humberside (S.),
DN17 3NU, UK

Contact: Mr Newbury
Tel: 0724 764606

Ilkeston & DMC & LCC
30 Lock Lane, Sandiacre,
Nottinghamshire,
NG10 5LB, UK

Contact: Mr R Woodward
Tel: 0602 393788

Ipswich MC & LCC
9 St Mary's Close,
Bramford, Ipswich, Suffolk,
IP8 4DL, UK

Contact: Mrs J Sago
Tel: 0473 742468

Isle Of Man Schoolboy Motor Cycling Ltd
Min-y-don, Ballaquayne
Park, Isle of Man, UK

Contact: Mr Kneen
Tel: 0624 842582

Islwyn MCC
14 Caernarvon Place,
Grove Place, Blackwood,
Gwent, NP2 1DB, UK

Contact: Mr T Ruck
Tel: 0495 223272

Kings Norton MCC
25 Cherhill Covert, Mony-
hull, Birmingham, Midlands
(W.), B14 5YB, UK

Contact: Mr F Dorrell

Lanarkshire MCC
11 Horsburgh Grove,
Balerno, Lothian,
EH14 7BU, UK

Contact: Mrs D.Stratford
Tel: 031 449 5718

Club trials once a month, classes
for all with special sections for
school boys/girls. For more
information please phone:
031 449 5718.

Leamington Victory MC & LCC
12 Dadglow Road, Bishops
Itchingham, Warwickshire,
CV33 0TG, UK

Contact: Mr A Halford
Tel: 0926 613202

Leicester Query MC
14 Anstey lane, Thurcaston,
Leicestershire, LE7 7JA, UK

Contact: Mr S Worth
Tel: 0533 363196

Lincoln MC & LCC Ltd
42 Baildon Crescent, North
Hykeham, Lincolnshire,
LN6 8HS, UK

Contact: Mr G Swaby

Littleport & DMC & LCC
14 North Street, Frecken-
ham, Bury St Edmunds,
Suffolk, IP28 8HY, UK

Contact: Mr P Gammon
Tel: 0638 721046

Llangollen & DMCC
2 Penllyn, Nant Parc,
Johnstown, Wrexham,
Clwyd, LL14 1YG, UK

Contact: Mr J Price
Tel: 0978 842142

Loughborough & DMCC
1 Meadow Crescent, Castle
Donington, Derbyshire,
DE7 2LX, UK

Contact: Mr J Harrad
Tel: 0332 810160

Lowestoft Invaders MCC
58 Dell Road, Oulton Broad, Lowestoft, Suffolk, NR33 9NS, UK

Contact: Mr R Greengrass
Tel: 0502 563566

Macclesfield Trials
3 Jackson's Brow, Pott Shirgley, Macclesfield, Cheshire, UK

Contact: Mr J McDonald
Tel: 0625 73633

Manby Showground
Sunny Oak, Little Cawthorpe, Louth, Lincolnshire, LN11 8ND, UK

Contact: James Tointon
Tel: 0507 604375
Fax: 0507 604092

Off road practice ground with specialist tracks for quads, pilots, moto-x, enduro, trials and 4 x 4. Test sessions must be prebooked by phone Monday to Saturday.

Manchester 17 MCC
2 Kinder Close, Glossop, Derbyshire, SK13 9UW, UK

Contact: Mr P Priestly
Tel: 0457 868469

Mansfield Maun MCC
Hillcroft, New Mill Lane, Forest Town, Mansfield, Nottinghamshire, NG19 OHH, UK

Contact: Mr G Morris
Tel: 0623 26688

Market Drayton MC & LCC
162 Shrewsbury Road, Market Drayton, Shropshire, UK

Contact: Mrs T Wycherley
Tel: 0630 3578

Merthyr Motor Club
Imperial Hotel, High Street, Merthyr Tydfil, Glamorgan, Mid-, UK

Contact: Mr B Marsh

Middlesbrough & DMC
165 Flatts Lane, Normanby, Middlesbrough, Cleveland, TS6 OPP, UK

Contact: Mr I Smith
Tel: 0642 454474

Midland Classic MCC
43 Sandringham Road, Sandiacre, Nottinghamshire, NG10 5LD, UK

Contact: Mrs W Silk
Tel: 0602 397601

Minsterley MC
12 Eskdale Road, Telford Estate, Shrewsbury, Shropshire, UK

Contact: Mrs M Bowdler
Tel: 0743 61348

Motorcycle Seatworks
366 Woodside Road, Wyke, Bradford, Yorkshire (W.), BD12 8HT, UK

Tel: 0274 604672

Manufacturers of every type of motorcycle seat cover, specialists in moto-cross. Appointment always necessary.

National Motorcycle Centre Trials Club
Park Promotions Ltd, The Post House, Ham Hill, Powick, Hereford & Worcester, WR2 4RD, UK

Contact: Mr T Matthews
Tel: 0905 830459

Norfolk & Suffolk Jnr MCC
Mill House, Stone Street, Crowfield, Ipswich, Suffolk, IP6 9SZ, UK

Contact: Mrs J Gibbons
Tel: 044 979 397

North Cornwall MC
14 Delaware Court, Delaware Road, Gunnislake, Cornwall, UK

Contact: Mr A Brockington
Tel: 0822 833288

North Derbyshire Youth MCC
542 Kedleston Road, Allestree, Derbyshire, DE3 2NG, UK

Contact: Mr A Nixon
Tel: 0332 559813

Northallerton & DMC
6 Wensley Road, Romanby, Northallerton, Yorkshire (N.), UK

Contact: Mrs Y Kirk
Tel: 0609 780105

Norwich Viking MCC
10 Willow Close, Lingwood, Norwich, Norfolk, NR13 4BT, UK

Contact: Mr A Hay
Tel: 0603 713017

Oswestry & DMC
Sherbrooke, Middleton Road, Oswestry, Shropshire, UK

Contact: Mr W H Jones
Tel: 0691 656471

Pathfinders & Derby MC Ltd
15 Westhall Road, Mickleover, Derbyshire, DE3 5PA, UK

Contact: Brian Tuxford
Tel: 0332 516861

Organise 10 trials per year on our own trials land near Derby. Regularly the East Midland centre's champion trials club. We cater for all ages.

Pegasus (Grantham) MC & CC Ltd
72 Barrowby Road, Grantham, Lincon, NG31 8AF, UK

Contact: Mr T Harris
Tel: 0476 65311

Pendennis MC & LCC
38 Lanner Hill, Lanner, Redruth, Cornwall, UK

Contact: Mrs S Pooley
Tel: 0209 215543

Perth & District MC
Flat 4, 4 Kirkhall Road, Almondbank, Tayside, PH1 3LD, UK

Contact: George Peterson
Tel: 0738 83877

Perth & District Motor Club is a trials-orientated club which currently runs three national Scottish trials per year.

Peterborough MCC
218 Broadway, Yaxley, Peterborough, Cambridgeshire, PE7 3NR, UK

Contact: Mrs L Rushbrook
Tel: 0733 240071

Phil Ayliff Products Ltd
25 Alliance Close, Attleborough Fields, Nuneaton, Warwickshire, CV11 6SD, UK

Contact: Trevor Ayliff
Tel: 0203 343741
Fax: 0203 641247

Manufacturer of Dunlopad sintered metal brake pads.

Pickering & District Motor Club
4 Brockfield Road, Huntingdon Road, York, Yorkshire (N.), YO3 9DZ, UK

Contact: Mr D A Brown
Tel: 0904 622274

Pickering and District Motor Club promotes motorcycle grass track, motocross and trials. The club has been in existence since 1950.

Ponthir British MCC
Heol Fawr Cottage, Rogerstone, Gwent, NP1 9GR, UK

Contact: Mr B D Gay

Potters MCSC
Hunters Croft, 88 Ash Bank Road, Bucknall, Stoke on Trent, Staffordshire, ST2 9DU, UK

Contact: Mr J Austin

Redditch MCC
Vaynor Drive, Headless Cross, Redditch, Hereford & Worcester, UK

Contact: Mr P J Smith
Tel: 0527 43265

Retford & DMCC Ltd
4 West Carr Road, Retford, Nottinghamshire, DN22 7NN, UK

Contact: Mr A Lane
Tel: 0777 701103

Rhondda Motor Club
310 Brithweurydd Road, Trealaw, Rhondda, Glamorgan, Mid-, UK

Contact: Mr K Richards
Tel: 0443 432434

Rugby MCC
10 Wordsworth Road, Rugby, Warwickshire, CV22 3HY, UK

Contact: Mr P Thornhill
Tel: 0788 813606

Salop MC
8 Oakwood Drive, Heath Farm, Shrewsbury, Shropshire, SY1 3EA, UK

Contact: Mr A Johnson

Scarborough MC
20 Broadlands Drive, East Ayton, Scarborough, Yorkshire (N.), YO13 9ET, UK

Contact: Mrs E Race
Tel: 0723 863987

Scottish Auto-Cycle Association
Block 2, Unit 6, Whiteside Ind. Estate, Bathgate, Lothian, EH48 2RX, UK

Governing body of motor cycle sport in Scotland.

Scunthorpe MCC
44 Pembroke Avenue, Bottesford, Scunthorpe, Humberside (S.), DN16 3LN, UK

Contact: Mr G Goddard
Tel: 0724 850468

Severn Valley Grass Track Club
14 St Davids Close, Lickhill Road North, Stourport-on-Severn, Hereford & Worcester, DY13 8RZ, UK

Contact: Mr J Doughty
Tel: 02993 4922

Solihull MCC
8 Sidenhall Close, Shirley, Solihull, Midlands (W.), B90 2QD, UK

Contact: Mr K McCoy
Tel: 021 745 3220

South Liverpool MC
81 Newton Road, Billinge, Wigan, Lancashire, WN5 7LB, UK

Contact: Mr F E Anderton
Tel: 0744 892422

South Shropshire MCC
Brooklyn Cottage, Batch Valley, All Stretton, Shropshire, UK

Contact: Mr J R Evans

Southern MCC Ltd
8 Birchleigh Close, Onchan, Isle of Man, UK

Contact: Mrs B Jones
Tel: 0624 622408

Spalding & Tongue End AC
45 Tanglewood, Werrington, Peterborough, Cambridgeshire, PE4 5DH, UK

Contact: Mrs S Cotton
Tel: 0733 74249

Stevens & Stevens Trials Centre
Unit 43, Blue Chalet Ind. Park, London Road, West Kingsdown, Kent, TN15 6BQ, UK
Tel: 0474 854265
Fax: 0474 854032

Trials motorcycles, spares, accessories & clothing. Advice on joining clubs (adults & juniors).

Stourbourne (Haverhill) MCC
79 Nothfield Park, Soham, Cambridgeshire, UK
Contact: Mrs K Edson
Tel: 0353 722681

Stowmarket & DMCC
25 Quinton Road, Needham Market, Suffolk, IP6 8BP, UK

Contact: Mrs V Hearn
Tel: 0449 721042

Sudbury MCC
14 Constable Road, Sudbury, Suffolk, CO10 6UG, UK

Contact: Mr M J Edwards
Tel: 0787 77033

Surrey Cycles
Surrey House, High Street, Cranleigh, Surrey, GU6 8RL, UK

Tel: 0483 272328

Amal-Mki Concentric carburettors and spares.

Sutton Falcons MC
42 Glebe Street, Annesley, Woodhouse, Nottinghamshire, NG17 9HD, UK

Contact: Mr P Flanagan
Tel: 0623 759032

Thirsk MC
2 Herriott Way, Thirsk, Yorkshire (N.), UK

Contact: Mr A Kendal
Tel: 0845 725937

Triangle (Ipswich MCC)
18 Monton Rise, Ipswich, Suffolk, IP2 9QQ, UK

Contact: Mr R Sayer
Tel: 0473 687698

Venhill Engineering Ltd
21 Ranmore Road, Dorking, Surrey, RH4 1HE, UK

Tel: 0306 885111
Fax: 0306-740535

Manufacturers of Nylocable and Featherlite control cables, Powerhose high performance stainless braided brake hoses. Distributors of Magura controls, Ariete and Buzzetti accessories and titanium and aluminium fasteners.

Vintage MCC (IOM Region)
19 The Crescent, Port-e-Chee Avenue, Douglas, Isle Of Man, UK

Contact: Mr K Teare
Tel: 0624 673618

West Cornwall MC
Oakland Cottage, Paul, Penzance, Cornwall, UK

Contact: Mr B Ellis
Tel: 073 673529

West Essex MCC
8 Hubbards Close, Hornchurch, Essex, RM11 3DH, UK

Contact: Mr C Crowder
Tel: 0787 473913

Whitchurch MCC
6 Chapel Crescent, Darliston, Whitchurch, Shropshire, UK

Contact: Mr C G Mellor
Tel: 0948 840143

Winsford & DMC
35 Shores Green Drive, Wincham, Northwich, Cheshire, CW9 6EE, UK

Contact: Mr D Buckley
Tel: 0565 733371

Wrexham MC
2 Wirral View, Penarlag, Hawarden, Clwyd, CH5 3ET, UK

Contact: Mr D Lovell
Tel: 0244 533342

Wymondham & DMC
10 Malton Court, Hethersett, Norwich, Norfolk, NR9 3HE, UK

Tel: 0603 810934

INDEXES

Index by alphabetical order

Renthal Ltd

Rock Oil Company

Rockhall Auto Electrics

Scootering

Shell Oils

Shoshoni Clothing

Speedway Riders Association

Streetfighters

Tommy Robb (Motorcycles) Ltd

Vintage Motor Scooter Club

Voc Spares Company

Wilson

Winsford & DMC

Cleveland

Front Line Motorsports

Guisborough & DMC Ltd

Middlesbrough & DMC

Northsport Motorcycles

Tillstons Ltd

Wire Wheels Services

Clwyd

A & D Motorcycles

Baglux UK

Colwyn Bay Motorcycles

Colwyn MCC

Denbigh & Mold MCC

Llangollen & DMCC

Mid Cheshire MXC

Motorcycle Club of Cheshire

Tony Hayward

Woods Motorcycles

Wrexham MC

Co. Antrim

BMF

East End Motorcycles

Hursts Auto Complex

J Morgan

R.F.Linton & Son

Co. Down

Bartel Alloy Tanks

G.S.Motorcycles

Norman Watt Motorcycles

Co. Dublin, Eire

Craigie Competition MC

Ireland Dublin Harley-Davidson

Kawasaki Distributors Ltd

Co. Durham

Dave Mackenzie

Scott Leathers International

Weardale Off Road Centre

Co. Wicklow

Timeless Distributors Ltd

Connecticutt, USA

Clubman Racing Accessories

Cornwall

BMF

British Motor Cycle Owners Club

British Sporting Sidecar Association

Callington Kawasaki

Cambourne-Redruth & DMC & LCC

Camel Vale MCC

Camelford Bike Bits

Cornwall Moto Cross Club

Cornwall Solo Grass Track Club

North Cornwall MC

Pendennis MC & LCC

Trail Riders Fellowship

West Cornwall MC

Cumbria

ALAPAT

Bill Brown's Motorcycle Centre

Bob Jackson

Crooks Suzuki Ltd

Jack Horseman Motorcycles

Kawasaki Carlisle Ltd

Keith Benton

Lakeland Motor Museum

Lloyd Lifestyle Ltd

Phil Cotton Classic Bikes

Starters Driving Centre

Derbyshire

Automotive Manufacturers Racing Assoc.

Bikers Gear Box

Clarks Motorcycles

Clay Cross Kawasaki

Colin Lomax Motorcycles

Crankshaft Specialists

Crusader Leathers Ltd

Darley Moor MRRC Ltd

Derby Phoenix MCC

Derwent MCC

Donington 100 MCC Ltd

Donington Race School

Granby Motors (Ilkeston) Ltd

Greeves Riders Association

Jim Matthews

Loughborough & DMCC

Manchester 17 MCC

North Derbyshire Youth MCC

Pathfinders & Derby MC Ltd

Paul Hunt Motorcycles

Powerslide Bikes

Premier Plates

Rawhide

Roy Pidcock Motorcycles

SEP

Silkolene Lubricants PLC

Summerfield Engineering Limited

The Paint Studio

VMCC Regalia

Velocette Heritage

Wilemans Motors

Youth Motorcycle Sport Associat on

Devonshire

BKS Leather

Bridge Garage (Exeter) Ltd

Bridge Motorcycle World

Bridge South West Motorcycle Training

Combe Martin Motorcycle Collection

Devon Rim Company

Doug Richardson

Exe Bike Breakers

Exeter Speedway

Federation of European Motorcyclists

GT Motorcycles

Goodridge (UK) Ltd

Harley-Davidson Riders Club

Hobbs Sport Earls Ltd

Independent Ignition Supplies

International Motorcyclists Tour Club

Just Bikers

Lambretta Museum

Lambretta Preservation Society

Lambrook Tyres Ltd

MPS

Mid-Devon Motorcycles

North Devon British Motorcycle Owners

PGH Motorcycles

Stators

Terry Hobbs Motorcycles

Totnes Motor Museum

Tracy Tools (BB) Ltd

Dorset

Acrybre Products

Armours

Association of Motorcyclists Against Discrimination

Barclay Motorcycles

Barclays Business Systems

Bike Monsters

Bike Tech (BMS Developments)

Britbits

CTG Racing Ltd

CW Motorcycles

Cagiva Three Cross (Imports) Ltd

Crescent Motorcycle Co.

Crescent Suzuki Bournemouth

Crescent of Bournemouth

Dorset Dirt Bikes

Francis Barnet Owners Club

House of Kolor

Le Velo Club Limited

Nevis Marketing Ltd

Poole Speedway

Renntec

Three Cross Motorcycles Ltd

Tri-Supply

Triumph Basket Cases

Wheeler Racing

Dumfries & Galloway

Ava Europa Ltd

Galloway M.C.C.

Rhins Motorcycle Club

The Tank Shop

Top Tek International Ltd

Dyfed

Bryn Haulwen

Garland & Griffiths

SRM Engineering

Essex

Arena Essex Speedway
B Perry
B Wybrow M/C Ltd
BMF
Braintree & DMCC
Battlesbridge Motorcycle Museum
Bill Roberts Race Fittings
Bits & Pieces
Brentwood Cycles
CES Ltd
Chadwell Motorcycles Of Essex
Chelmsford & DAC
Chris Applebee
Coulson Engineering Services Ltd
DKW Rotary Owners Club
Dunmow & DMCC Ltd
East Anglian Schoolboy Scrambling Club
Eastern Centre Marshals Club
Eastern Four Stroke Association
Eastern Sporting Sidecar Association
Eddy Grimstead Ltd
Essex Enduro Club
Essex Schoolboy/Adult Trials Club
Essex Trials Group
Essex West Road
Federation of Sidecar Clubs
Frank Webber
Gilera Information Exchange
Guy Fensome T.S.
Halstead & DMCC
Harwich MCC
Heritage Harley-Davidson
Ilford MC & LCC
Jim Aim Motorcycles
John E Vines Ltd
John Pease Motorcycles
Johns of Romford
Julian Soper Motorcycles
Kelvedon Hatch Trials Club
Kendall & Pitt
Leo Yuill
National Auto-Cycle and Cyclemotor Club
Phillips Transfers Ltd
Precision Engineering Services
R.W.Parkinson & Son
Scott Owners Club
South Essex Motorcycles
Southend & DMCC
Speedway Association
Swann & Moore (Assessors) Ltd

Tec-Nick
Technique Tyres Ltd
The Finishing Touch
Triple Cycles
Triumph Owners MCC
West Essex MCC

Fife

Alan Dufus M/C
Dunfermline & District M.C.C.
GWD M/C Spares
Glenrothes Youth M.C.C Ltd
Kinross & District MCC Ltd
Kirkcaldy & District M.C.Ltd
Knockhill M.C.R.C.Ltd

France

Simpson Mecanique
Simpson Mecanique Ducati

Germany

Ogri MCC
International Motorcycles

Glamorgan (S.)

BMF
British Bike Magazine
British Motorcycle Engineering
Cardiff Motor Club
Motorcycle Mart
R.C.M.
Richards Brothers
Robert Bevan & Son

Glamorgan (W.)

Dick Weekes
JT's Motorcycles
Kick Start Motorcycles

Glamorgan, Mid-

Caerphilly & DMCC
East Wales Moto Cross Club
Gilfach Triangle Motor Club
Leslie Griffiths Motors
Merthyr Motor Club
Mid Glamorgan MC
Motorcycle World
Rhondda Motor Club
Two Wheel Services ltd
V.M.C.C. Transfer Service

Gloucestershire

Anglo German Tools
BSA Company Limited
BVM Moto
BVM Trialsport
British Schoolboy Motorcycle Assoc.
Claremont Motorcycles
Cotswold Motor Museum
Fred Cheshire Motorcycles Ltd
Gloucester Kawasaki Centre
Gloucestershire AMS Motorcycles
Hongdu - Anglo Hazet German Tools Ltd
MZ Motorcycles GB Ltd
Manx Grand Prix Riders Association
Michael Freeman Motors
Racespec
Tyre Sales Motorist Centre
Veteran Vespa Club
Watsonian-Squire Limited

Grampian

Aberdeen & District Moto Cross Club
Bon Accord MCC
Cougar Customs
Grampian M.C.C.
Grampian Transport Museum
Kawasaki Aberdeen
Shirlaws Motorcycles
VMC North East Section

Guernsey, CI

British Motor Cycle Club
Nady/Powerline Limited
St Peter Port Garages Ltd

Gwent

Abergavenny Auto Club
British Sporting Sidecar Association
Caerleon & District Motor Sports Club
Ebbw Vale MCC
G.U.Products
Gwent Schoolboy Trials Club
Islwyn MCC
MXB Motocross
Manmoel Motocross Practice Track
Nantyglo & DMCC
Ponthir British MCC
Pontypool & DMCC
South Wales Superbikes
Taffspeed
Welsh Sprint Society

Gwynedd

Conway & DMCC
Morgan Three Wheeler Club Ltd

Hampshire

AJS Motorcycles
Ash Vale Motorcycles
Automobile Association
BMF
BSA Owners Club
Bifax International Ltd
Bike Business
CBX Riders Club (UK)
Colin French
Collins Tyre Service
Compensation Direct
Des Helyar Motorcycles
Egli Vincent Owners Club
Fareham Electroplating Co.Ltd
Fedn.of British Police Motorcycle Clubs
Hill Marketing Ltd
Honda Owners Club (GB)
JB's
MZ Motorcycles GB Ltd
Military Vehicle Trust
Motorcycle City
Motorcycle City Sales
National Motor Museum
Norman White Norton
Norton Owners Club
Pedal & Motor Limited
R.D.Cox & Son
Rafferty Newman
Ray Dentith
Rob Willsher Motorcycles
Rob Willsher Suzuki Spares
Royal Navy MCC
Rye's of Southampton
Sammy Miller Museum
Sidecar Register
Southern Motorcycles
T.R.F. Bulletin
Thruxton Motorcycles
Tran Am Ltd
Two Wheels
Vintage Japanese MCC

Leicestershire

BMF Rally
Bruce Main Smith & Co. Ltd
Charnwood Classic Restorations
Clive Castledine Motor Cycles
Clubmans Racing Club (EM)
Cross Street Kawasaki
Donnington Supporters Club
Drayton Croft M/C
Earl Shilton Trials Club
Eddy Grew
Exports-Imports
Ferodo
KTM Motorcycles
Leicester Query MC
Len Manchester
M/C Accessories
Marcol Motorcycles (Leicester)
Market Harborough British M/C
Club
Martyn's Motorcycles
Nourish Racing Engine Company
P&D Consultants
R J Motorcycles
Redcar Motorcycles
Stanford Hall Motorcycle Museum
Sunbeam Owners Fellowship
Supreme Motorcycles
Triumph Designs Limited
Triumph Motorcycles Ltd
Twiggers Motorcycles
UK Motorcycle Exports
Vincent HRD Series A Owners Club
Weeden Restoration
Windy Corner

Lincolnshire

A.P.S. Sales
Alloy Polishing Service
B & C Express Products
C.H.Biggadyke M/C
C.S.M.A. (East Midlands)
Dirt Bike Rider
Forgotten Racing Club
Francis Motors
Geeson Bros.Motorcycle Museum
Italia Classics
Jack Machin M/C
Lincoln MC & LCC Ltd
Lintex Gleave Ltd
Louth & DMCC
M R Holland (Distributors) Ltd
Manby Showground

Mike's Bike Spares
Mill House Books
Moto-Bins
Nova
Pegasus(Grantham)MC & CC Ltd
Phoenix Supplies
R A Wilson Motorcycles
Skegness & DMCC
TRM Racing
Wainfleet & District Sporting MCC
Webbs of Lincoln

London

59 Club Classic Section
Ace Classics London
Archway Project
Army Shop Ltd
BMF
Bat Motorcycles
Bikerama Ltd
Bob Porecha
Boyer
British Bike Insurance
CHL Engineering Ltd
Connoisseur Carbooks
Cosmopolitan Motors Ltd
Cylinder Head Shop
D J Allwood
David Levene & Co
Design Museum
Devitt Ltd
E A Grimstead & Sons Ltd
F H Warr & Sons
Fast Bikes
Fellowship of Christian
Motorcyclists
Frontiers Motorcycles Ltd
Geoff Dodkin
Going Places
Hagon Products Ltd
Hamiltons Motorcycle Centre
Hamrax Motors Ltd
Honda Gold Wing Owners Club
Honda Motor Europe Ltd
Honda UK Ltd
Ian Melrose
Ilford Amateur MC
Institute of Motorcycling
Jack Nice Motorcycles
Joe Francis Motors Ltd
Kawasaki In London
Kinetic Art Paintshop
Lexport (Sales & Service) Ltd

London Biker
London Suzuki Centre
MW Leathers
McLaines
Mettbikes Breakers
Mitchell & Partners
Mobil Oil Co.Ltd
Motorcycle News
Motorcycle City
Motorcycle Retailers Association
Motorcycle Sport
NGK Spark Plugs (UK) Ltd
National Deaf Motorcycle Club
National Motorcycle Council
North Weald MCC
Parks of Lewisham
Performoto
Peter Bond
R. Judd Ltd
RAC Motoring Services
RAP International Superbikes
Raleigh Safety 7 & Early Reliant
Owners
Reg Allen
Rickman Owners Club
Riders Union
Science Museum
Screencraft Ltd
Silver Machine
Slocombes
Sondel Sports
Spares (GB)
Sporting Cycles
Streetbike Drag Club
Sumo Bikes
Tony Bairstow
Tregunna
Triumph In London
Westminster Communications
Group
Wheelpower Bike Centre
Women's Int. Motorcycle Assoc.
Yamaha Riders Association

Lothian

Alvins Motorcycles
Better Bikes Ltd
C J Wilson
Carrick Motors
East Lothian Road & Trail Club
Edinburgh & District M.C.Ltd
Edinburgh Southern M.C.
Edinburgh Speedway
Ernie Page Motors Ltd

Lanarkshire M.C.C.
Myreton Motor Museum
Royal Museum of Scotland
Scottish Auto-Cycle Association
Scottish Vintage Racing Club
West Pier Motorcycles

Manchester, Gt.

Anglo-Scot Abrasives
Aprilia Moto UK
BMF
Belle Vue Speedway
Boxer-K Motorcycle Services Co
Classic Brake Services
DOT Owners Club
Doug Hacking Motorcycles
Dunstall Owners Club
Italsport
Jack Bottomley's
KAIS
Manchester Eagle MCC
National Handicapped Motorcyclists
Queens Park Motors Ltd
Robinsons of Rochdale
Scientific Coatings
Thunderbird Classics
Westhoughton Classic & Modern
MCC
Worsley Motorcycles

Merseyside

BSS Motorcycles
Bike Spurz
Bill Smith Motors
Cheshire Youth Trials Club
Davida (UK) Ltd
Harley-Davidson of Southport
Hawkshaw Motorcycles Ltd
K Southport Superbikes
Kirwans Solicitors
Lambretta Club GB
Littlewoods Organisation
Mal Kirwan
R.A.V.Engineering
Southsport Superbikes
St Lukes Motorcycles
Supreme Visors Ltd
TT Tyres

Middlesex

Advanced Motorcycle Systems

Blays

Border MCC

CBS (Whitton) Ltd

Colin Collins Motorcycles

Continental Tyre & Rubber Group Ltd

Daytona

Delta Clothing

Enfield & District Veteran Vehicle Trust

H E Webber & Sons

Jack Lilley Ltd

Miles Engineering Company

Moto Guzzi Club (GB)

Motorcycle City

Motorcycle City Sales

P & M Motorcycles

Renham Motorcycles

Roebuck Motorcycles

Rolling Thunder UK

Spartan Kawasaki Owners Club

Trail Bike Fun Enduro Club

Trident Engineering

Midlands (W.)

A.J.Whitehouse

Advanced Motorcyclists Assoc.UK

Alpha

Amal Concentric Carburettors

Amateur Motorcycle Association

Association of Independent Motorcyclists

Association of Road Race Clubs

Baby Biker

Bennetts Insurance Brokers

Birmingham MCC

Birmingham Motorcycles Ltd

Birmingham Museum of Science

Bob Heath Visors Ltd

Bon Accord Leisure Ltd

Bridgestone/Firestone UK Ltd

C.S.M.A. Ltd (Midland)

Central Wheel Components Ltd

Cetem Polishing Kits

Coleman Body Repairs

County Border MCC

Cradley Heath Kawasaki

Cradley Heath Speedway

D.M.W. Motor Cycles

DJF Metal Polishing

Dirtwheels of Coventry Ltd

Drayton Croft M/C

Dudley & District MCC

Dynamic Ltd

Early Stocks RC

East Midlands Racing Association

Elbe Motocross

Epoxy Powder Coatings Ltd

Express Bike Spares

Feridax (1957) Ltd

Footman James & Co Ltd

Graham Engineering

Grimeca

Harlgo Limited

Harrison Scrambling Track

Heinkel/Trojan Owners/Enthusiasts Club

Hitchcock's Motorcycles

Hughie Hancox

Kings Norton MCC

Le Velo Club

Magnews

Magnum

Maico Spares

Marston Sunbeam Register

Middle England Scooter Association

Midland (Moto) Rumi Club

Midland Drag RA

Motad International Limited

Moto Imperial Supplies

Motor Cycle Industry Assoc. of GB Ltd.

Motor Vehicle Imports Ltd

Motorcycle Action Group

Motorcycle Association of GB

Motorcycle Equipment Ltd

Museum of British Road Transport

NCK Racing

National Motorcycle Museum

Off Road Riders Scooter Association

Peatol Machine Tools

Penrite Oil

Power Torque Engineering Ltd

QB Motorcycles

R.K.Leighton

Richard Hughes

Rochdale Owners Club

Route 66

Royal Enfield/Hitchcocks Motorcycles

S P Tyres (UK) Ltd

SPC Bearings Ltd

Sandwell Heathens

Sherwood Garage

Silks Solicitors

Solihull MCC

Solihull Motorcycles

Sonic Communications (Int) Ltd

Speedway Motorcycles

Speedway Motors

Street Bike

Streetbike (Dudley)

T.T.Restorations

TT Supporters Club

Tattoo Factory

The Bonneville Shop

The Museum of Science & Industry

Top Gear

Triskill Ltd

Two Wheel Tyre Centre

Vale - Onslow

Veteran Speedway Riders Association

W.E.Wassell

Wolverhampton Speedway

Midlothian

Edinburgh St George M.C.

Lothian & Borders C&VMCC

Melville M.C. (Scot) Ltd

Minnesota, USA

ACS International

Norfolk

Barber Engineering

British Formula Racing Club

Caister Castle Motor Museum

Caravan Camping & Leisure

Denver Motorcycles

Diss & District MC & LCC

Ducati Owners Club

Freewheel UK Ltd

Hewy Design Paintwork

Jawa - Skoda (GB) Ltd

Jaybee M/C Ltd

Kings Lynn & District MX Club

Kings Lynn Speedway

M & S Weatherley

Mike Bavin Motorcycles

Motorcycle Clothing Centre

Norfolk & Suffolk Group

Norwich & Dist SB/AD MXC

Norwich Union Fire Insurance

Norwich Viking MCC

PSP Engineering Services

Skoda (GB) Ltd

Staccato Ducati

Steve Bullock

Tinklers Motorcycles

Watton DMC & CC

Wymondham & DMC

Northamptonshire

Anchor Kawasaki Centre

BMF

BMF Midlands Region

BSA Bantam Racing Club

Bike Gear

Bike Magazine

Chassis Dynamics

Classic Mechanics

Corby Kawasaki Centre

Frank Thomas Ltd

Freeman Automotive UK Ltd

Harley-Davidson UK Ltd

Italian Motorcycle Owners Club

KME Racing

Kawasaki Spares

Mick Hemmings

Midland M/C

Moto Cinelli

Motocross Practice Site

Motorcycle Dealer

Neil Young Motorcycles

New Era MCC

Phil's British Bike Bits

Prison Motorcycle Brotherhood UK

Rockingham Forest Tourism Association

Ron Jenkins

Shadowfax Engineering

Trident & Rocket 3 Owners Club

Northumberland

North East British Motorcycle

North East Motorcycle Racing Club

Robin Watson Signs

Rotascraft Engineering

Nottinghamshire
A.Gagg & Sons
BSSA (East Midland)
Benelli (Brembo Club International)
Colin Gregory M/Cs
D.Wilkinson
East Midland Scooter Association
Euro-Helmets Ltd
Forgotten Racing Club
Fox Motorcycles
Fox's of Worksop
Gay Bikers MCC
Granby Motors
Ilkeston & DMC & LCC
JR Technical Publications Ltd
Kawasaki Triples Club
Long Eaton Enamellers
Long Eaton Speedway
Mansfield Maun MCC
Marcol Motorcycles
Martin's Leathers
Midland Classic MCC
Moto Elite
Motorcycle Racing Sponsors Association
Motul Motor Oil (UK)
National Scooter Riders Association
National Scooter Sport Association
RIGP Finance Ltd
Retford & DMCC Ltd
Road Racing Monthly
Startline Accessories
Startline Motorcycle Accessories
Sutton Falcons MC
TMS
Triumph BSA
Velocette Owners

Oxfordshire
Abingdon Motorcycles
British Motorcycle Riders Club (Oxford)
Crowmarsh Classic Motorcycles
Holloway's - Auctioneers
Hughenden M40
International Christian Classic
J.W.Tennant-Eyles
Maico Owners Club
Oxford M/C Engineers Ltd
Oxford Products Ltd
Oxford Speedway
Stanford Auto Factors
T.Johnson Cables GB
T.W.Motorcycles

Perthshire
Vintage Motorcycle Club Ltd

Powys
Aberaman & DMCC
Custom Fasteners Ltd
Cwm Derw
David Jones Motorcycles
Exilla Leisurewear
Mid Wales Kawasaki Centre
Ron Kemp
Supersprox

Shropshire
Bantam Grasstrack Association
Fellowship of Historic Motorcycle Assocs
Market Drayton MC & LCC
Midland Motor Museum
Minsterley MC
Morris Lubricants
Oswestry & DMC
Oswestry Bicycle Museum
Peter Jones
Phoenix
Salop MC
Sam Pearce & Son
Shropshire Grass Track Club
South Shropshire MCC
Sports & Vintage Motors
Telford Motorcycle Breakers
Whitchurch MCC
Wylie & Holland

Somerset
Assoc. of Pioneer Motor Cyclists
Bryan Goss Motorcycles
Cheddar Motor & Transport Museum
Dragon Skin
Export Sales & Mail Spares
Haynes Motor Museum
J H Haynes & Co Ltd
Medium Link Ltd
National Hillclimb Association
Pat Watts Practice Track
Post Office Vehicle Club
Riders (Bridgewater) Ltd
Spellbound
Stibbs
Westland Motorcycle Club

Staffordshire
Bassets
Bob Price Classic Bikes
Brian Bennett
Burton Bike Bits
C.G.Chell Motorcycles
C.M.Smith
Cheadle (Staffs) AC
D H Autos
D&K
Fair Spares
Fine Thompson Ltd
Fred Barlow
G K Blair
GT/GTR Kawasaki Owners Club
Greens of Longton
Hemming & Wood Ltd
Jacksons Motorcycles
Kirby Rowbotham
MMCRC
MV Agusta Owners Club
MZ Riders Club
Motozone
NSU Register
Norton Motors
P.W. Ranger
PHD Metal Refurbishment
PW Ranger
Phoenix Distribution Ltd
Potters MCSC
Protectorl Ltd
Renold Chain Ltd
Rugeley Motorcycles
Sportbike
Suzuki Owners Club
Tamworth & District Classic MCC
Vintage Motor Cycle Club Ltd
Worldwide Norton Riders

Stirlingshire
VMC Stirling Castle Section

Strathclyde
Avon Valley M.C.C
Garnock Academy M.C.C.
Glasgow Speedway
Loch Fyne Motorcycle Club
Mickey Oates Motorcycles
Museum of Transport
North Harbour Motorcycles
P&J Powder Coatings
Ride On Motorcycles Ltd
Scotbike Ltd
Scott Oilers
Scottish Classic Racing M.C.C.
Stevenson & District M.C.C.
Strathclyde Trading Co.Ltd
Victor Devine & Co Ltd

Suffolk
500cc Sidecar Association
Andy Tiernan Classics Ltd
Ariel
Ariel Dragonfly
Baycover Ltd
Bickers Anglia (Access) Ltd
Bury St Edmunds & DMCC
C.J.Bowers Ltd
Castle (Colchester) MCC
Civil Service Motoring Association
David Silver Spares
Dragonfly Motorcycles
East Anglia Transport Museum
Easton Farm Park
Essex & Suffolk Border MCC
Faithfull Classics
Ipswich MC & LCC
Ipswich Speedway
Littleport & DMC & LCC
Lowestoft Invaders MCC
Newmarket MCC
Norfolk & Suffolk Jnr MCC
Orwell Cycles
Revettes Ltd
S.C.Engineering
Sevenhills Motorcross
Stowmarket & DMCC
Sudbury MCC
Suffolk Moto Parc
Triangle(Ipswich MCC)
Woodbridge & DMCC

Yorkshire (S.)
British Motorcycle Preservation
Society

Carnells

Cusworths Kawasaki

David Earnshaw

Dawson Harmsworth Ltd

Direct Tyres

FTW Magnetos & Dynamos

FTW Motorcycles

Pollards Motorcycles

Rainbow Motorcycles

STW Motorcycles

Sheffield Speedway

Staniforth Ltd

Yorkshire (W.)
ACE Electroplaters

Ashtech Limited

Association of Hill Climb Clubs

Automobilia Transport Museum

Autovalues Engineering

Bradford Ignition Services

Bradford Speedway

C & I Threading

C Wylde & Son Ltd

C.S.M.A.(East Yorks)

Castle M/C

Castleford M/C

Cobb & Jagger

Colin Appleyard Ltd

Earnshaws Ltd

End to End Endurance Club

Gizmo Marketing

Hobbsport Racing

Kawasaki Auotorama

Lee Bros

Mercer Skilled Crafts Ltd

Motorcycle Seatworks

Motorworks

National Breakdown Recovery

National Sprint Association

P & R Motorcycles

Padgetts (Batley) Ltd

Peter Watmough

Readspeed

The Harley Shop

Walker Engineering

York Suzuki Centre

ADVERTISERS